Soil Grown Tall

Gregory J. Retallack

Soil Grown Tall

The Epic Saga of Life from Earth

 Springer

Gregory J. Retallack
Department of Earth Sciences
University of Oregon
Eugene, OR, USA

ISBN 978-3-030-88741-4   ISBN 978-3-030-88739-1   (eBook)
https://doi.org/10.1007/978-3-030-88739-1

Cover photo: Diane Retallack on the Lint silt loam soil, a dark volcanic soil (Andisol), on the costal
terrace near Yachats, Oregon.

This Springer imprint is published by the registered company Springer Nature Switzerland AG
The registered company address is: Gewerbestrasse 11, 6330 Cham, Switzerland

Dedicated to Nicholas and the quest for understanding

Preface

Soil is often overlooked and underestimated. Yet all we are, and much of what we have, came from the soil and will return there when we are done. By recycling our dead and by purifying our water, soil saves us from disease. Soil is an important carbon sink and neutralizer of weathering acids, mitigating extreme greenhouse warmings of the geological past. Past cycles of life and climate have also been cycles of soil. How such a humble substance does so much is astounding and a topic to which these pages return again and again. But additional mysteries are explored in this personal quest to understand the long geological history of soil. How did it develop over the ages, this nurturer of life, this filter of air, this political resource? Did life evolve from soil or soil evolve from life? Were forests made by forest soils or forest soils made by forests? Are nations built from or merely on their soils? Did global change alter soils or did soils cause global change? These are difficult questions, but fortunately soils have a fossil record. Fossil soils or paleosols are now recognized to be abundant among sedimentary rocks. They are especially common in colorful red beds and in grimy coal measures. This book is dedicated to the ongoing task of deciphering this long fossil record of ancient landscapes and the ecosystems they supported.

Eugene, USA

Gregory J. Retallack

Acknowledgments

This book is a distillation of two decades of a rich and fulfilling career as a geology professor. It owes much to students and faculty of the University of Oregon and to research grants from the US National Science Foundation. They provided the means, encouragement, and stimulus to explore fundamental questions of soil evolution. Much of this book was written in tented camps in Antarctica and Kenya, providing evening diversion after days of fieldwork. I am particularly thankful to students and colleagues for advice and editorial help: Jonathan Wynn, Erick Bestland, Evelyn Krull, Helen Vallianatos, Scott Robinson, Shaun Norman, Lynn Soreghan, Christine Metzger, Carolyn Phillips, Jim Farlow, and Egbert Leigh. Last but not least, I thank my family, Diane, Nicholas, and Jeremy, for teaching me the importance of our garden.

Contents

List of Figures

List of Color Photos

1

Rainbow Rocks

Paleosols, or buried soils, are widespread in multicolored badlands, and provide evidence for climate and other surface conditions between episodes of sedimentation.

We are trained to see life as clean and soil as dirty, but life and soil are not so different. Soils are born and die during catastrophic events such as floods and landslides. Many lives are lost or established by such catastrophes. Between these end points, soils develop and grow by addition of clay and organic matter from maturing populations of plants. Soils have multiple metabolic pathways, not only of their contained multitudes of microbes, but their own distinctive chemical reactions, such as weathering of feldspar to clay. Soils reproduce when wind or water scatters their clay and dust across the landscape. Thus soils can be viewed not only as nurturers of life, but as a starter system for life itself, a form of protolife. Back to the beginning the line between soil and life may have been fine enough to have been imperceptible, as the tiniest nooks and crannies of soil continued their chemical reactions, not in open intergranular spaces, but within the first isolating membranes of the earliest cells. Could it be that, after all, we are but soil grown tall?

Soils are at the nexus of life and air, and fossil soils have changed substantially over geological time as life and air evolved. Thus this book runs backward into deep time to develop an appreciation of the coevolution of life and soil over the ages that followed its origin in and as soil (Fig. 1.1). Nevertheless, the concept of a fossilized soil needs some explaining, as was made clear by Vladimir Nabokov. The opening page of his novel *Lolita* mentions paleopedology, the study of fossil soils, as the epitome of an obscure scientific interest of Humbert Humbert. So allow me to explain fossil soils, and why I

© The Author(s), under exclusive license to Springer Nature
Switzerland AG 2022
G. J. Retallack, *Soil Grown Tall*,
https://doi.org/10.1007/978-3-030-88739-1_1

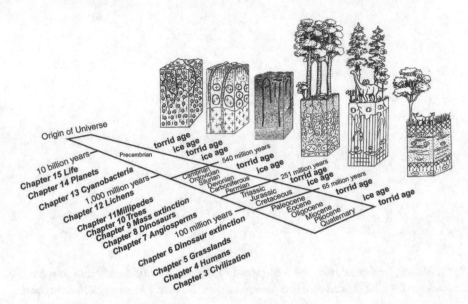

Fig. 1.1 A long view of soils and their biota backward into geological time, showing major events in atmospheric history and an outline of the chapters of this book

devoted an academic career to the study of something regarded as obscure by Nabokov.

You have seen many fossil soils, even if you did not recognize them, in the rainbow-colored landscapes of western movies. The enduring appeal of westerns derives in part from the freedom of time and space that is everywhere apparent in the wide, open landscapes of the North American West. The Bighorn Basin of northwestern Wyoming, for example, is a dusty rangeland of sagebrush rimmed with furrowed badlands of crumbling clay. The badlands are "bad" because they defy the efforts of ranchers to use them for pasture or grain, or of pioneers to cross them in wagons. The untamed badlands are a law unto themselves, a metaphor for the Old West itself. The badlands clays are a riot of red, green, orange, brown, and yellow bands and splotches. The beauty and fascination of Wyoming badlands was well expressed by Christina, the four-year-old daughter of geologist Mary Kraus, who has returned to Wyoming year after year to study these colorful rock exposures. "Those are mommy's rainbow rocks!" Christina once exclaimed. Her remark echoes Navajo Native Americans, who call comparably colorful rocks of Capitol Reef, Utah, "land of the sleeping rainbow." Navajo elders sensed what modern science has confirmed. The badlands are a record of events and environments of the deep past. There is a common thread to these colorful rocks and others like them in Bryce Canyon National Park, the Badlands of South Dakota,

Color Photo 1.1 Early Cretaceous (112 million years), Alag (grey-purple Inceptisol), Boro (red Inceptisol), and Mulani (Entisol) paleosols in the Zhonggou Formation at Zanye Danxia Geopark, Gansu, China

Arizona's Grand Canyon, the Flaming Cliffs of Mongolia, Olduvai Gorge of Tanzania, the Ischigualasto badlands in Argentina, or the Flinders Ranges of South Australia (Color Photos 1.1, 1.2). They include buried landscapes of the past, represented by numerous fossil soils, or paleosols.

A paleosol is clay, silt, sand, gravel, or rock that was formed as a soil at the Earth's surface, and then was preserved, usually by burial, as a record of a former landscape and its life. Between times of flooding and volcanic eruption, grasses, trees, worms, and rabbits rooted and burrowed the freshly created ground and worked it into productive soil. The product of this concentrated biological work over long periods of time can include red, blocky clays, or white, hard nodules, or black seams of coal. Each of these materials form very distinct horizons in otherwise ordinary shales, sandstones and conglomerates. Life and soil have had a long and intimate relationship. When fossils are not preserved, paleosols may be the only records of past life.

The concept of paleosols can give surprising answers to what otherwise would be complete mysteries, and my own appreciation of paleosols grew slowly from childhood to adolescence. When I was a child, I lived in the dirt, not in front of a computer screen, and my mother often urged me to "Go play outside!" From the age of five I was fascinated with enigmatic, dark, red, rocklike lumps in brown soils near Sydney, Australia. What on earth

Color Photo 1.2 Oligocene (32 million years) Luca paleosols (red Alfisols) in the Big Basin member of the John Day Formation in the Painted Hills, central Oregon

were they? These dark red nodules and concretions are difficult to characterize succinctly because of their great variety. Their size ranges from that of peas to tennis balls. In shape, they vary from spheres to misshapen lumps, like small figurines that allow ample scope for childish imagination. Some are hard, heavy and polished. Others are hollow and partially filled with dusty, yellow ocher, a distinctive form called a paint-pot concretion (Fig. 1.2). Selected nodules were prizes of my childhood rock collection arranged on cotton wool in cigar boxes, or within the small compartments of boxes designed for fishing tackle.

The nodules and concretions are widespread around Sydney, as I found in moving between suburbs and schools, and that is one clue to their origin. It does not seem to matter whether the soil containing them is gray and sticky, brown and crumbly, or orange and hard. The hard, red lumps are in, on, and under these very different soils. As the soils vary, so do their vegetation, but soils under woodland, heath and meadow all contain similar nodules. Thus it seems unlikely that the nodules were formed by any particular kind of soil or vegetation. Nor does local bedrock seem to matter. They were not in any particular rock layer, nor along faults or joints. Each excavation for building foundations or roads in sandstone, shale, claystone or dolerite brings up a new crop of dark reddish lumps. Unlike their surrounding shales that weather to

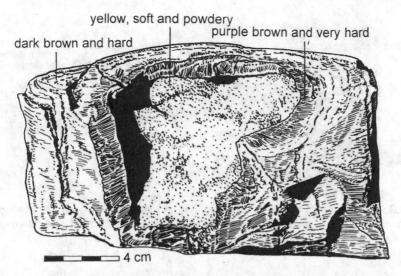

dark brown and hard

yellow, soft and powdery

purple brown and very hard

4 cm

Fig. 1.2   A paint-pot concretion found in a thick soil near Toongi, New South Wales, Australia, has a dark, hard, red, outer rind enclosing a soft, powdery yellow interior. These are fragments of an ancient surface soil, stained red and hard along the original soil cracks, but soft and yellow in the interior of the weathered block defined by the soil cracks

soil, the red nodules remain like the coagulated blood of a deep wound into the bosom of mother earth.

Additional clues to the origin of the red nodules became apparent during a childhood infatuation with archeology when I discovered that the red nodules were very old. In cliffs, the red lumps merge into sandstone bedrock, forming thick, wide ledges. In places, wind and dust have etched them from sandstone into grotesque shapes, like gargoyles on a Gothic cathedral. In such places they formed before the erosional cutting of the deep modern canyons, which have very small streams at the bottom and must have taken a long time to erode. The red nodular bands and bulges are cut by the carved grooves of ancient ceremonial grounds of the aboriginal peoples of Australia. The stains also are overlain by chains of arranged stones with the patina of long exposure at aboriginal ceremonial grounds, such as those in Kuringai Chase National Park, which has not been occupied by aboriginals for hundreds of years. Thus the red nodules and bands predate these ancient human carvings and constructions. The red stain passes in and out of soil and rock like spirits that meld time present and time past, an embodiment of the Dreamtime of the Aboriginals. They were yet another mystery of the inscrutable and timeless Australian landscape.

brown
soil

reddish brown
Triassic paleosols

red
Miocene
laterite

Fig. 1.3 Thick (30 ft.), red, nodular, Miocene (16 million year old) laterite on the beach at Long Reef, near Sydney, Australia, with a sequence of red paleosols of Triassic geological age (248 million years old) in the background

By the time I was admitted to university, I found a scientific explanation for these common red nodules and concretions. Ron Paton, one of my professors at Macquarie University, had worked extensively with similar materials in Africa and Southeast Asia. They are lateritic remnants of ancient soils, from 16 million years ago. Soils and life were different then from today's dry open woodlands of gum trees (*Eucalyptus*) and native "apple" (*Angophora*), with narrow, leathery leaves drooping in the intense summer heat. Back in the Miocene epoch some 16 million years ago, southeastern Australia was covered by cool, wet, jungles of fig trees (*Ficus*) and cabbage palms (*Livistona*), with soft, broad, spreading leaves. Some of these plants persist locally in isolated canyons of the sandstone plateaus, as a relict flora of this time of unusual warmth and humidity for this part of Australia. Roots of this lush vegetation of the past reached more deeply into the rock than the roots of plants today, and they were more effective at weathering the soil and mobilizing its nutrients, including iron. Dissolved in groundwater, iron came into contact with air and filled tiny pores in the rock and soil with reddish, iron oxide minerals, goethite and hematite. These patchy residues of deep weathering by past jungles are the red gargoyles and figurines that litter Sydney landscapes. Soils formed under eucalypt woodland of Sydney today, in contrast, are brown to yellow and clayey. The red nodules and concretions are from much older horizons of soils no longer forming, which in most places have been eroded away to create the current landscape of canyons and cliffs.

Ron Paton knew of complete weathering profiles of these ancient jungle soils preserved in a few areas around Sydney, for example, in road cuts near Frenchs Forest and Terry Hills, and on the beach at Long Reef (Fig. 1.3, Color Photos 1.3, 1.4). In places, the red nodular rock is tens of feet thick and forms

Color Photo 1.3 Middle Miocene lateritic Oxisol over Early Triassic (Spathian) Bald Hill Claystone with Long Reef paleosols (red Ultisols) in background cliff at Long Reef, New South Wales, Australia

Color Photo 1.4 Middle Miocene red laterite and white china clay under orange Oxisol, and dark brown Holocene (4 ka) Mollisol soil at Long Reef, New South Wales, Australia

a cliff above a white and red mottled clayey horizon of profoundly altered bedrock. The white clayey patches are the most weathered, and are now high-grade China clay. These clays are prized for fine pottery and known under the mineral name kaolinite, after a region in China famous for such pottery. These were soils much thicker and more deeply weathered than the brown to gray sandy surface soil of today's scrubby woodlands. The red nodular rock is called laterite, which comes from a Latin word for "brick". In India and Southeast Asia, laterite is dug from the ground as soft rectangular masses that harden into bricks when dried in the sun. Temples and walls of Angkor Wat near Siem Riep in Cambodia are made from bricks cut from ancient jungle soils. Sydney's red paleosols are also ancient, and also are fossil traces of past jungles. They have proven difficult to date, but associated pollen grains in surrounding sediments, and magnetic properties of the laterites has converged to indicate an age of middle Miocene, or some 16 million years old, for the lateritic paleosols of Sydney.

These lateritic paleosols were just the beginning of my personal scientific odyssey with paleosols, soon overshadowed by my discovery of other pale-osols, geologically much more ancient, in the sea cliffs flanking Sydney's famous surfing beaches. Our family camped for several summers at Palm Beach, named for cabbage palms (*Livistona*), which are relics in sheltered valleys of the Miocene jungles that have given way to eucalypt woodlands of today. The beaches at Palm Beach, Avalon, Bilgola, and Narrabeen were favorite teenage haunts of mine. When the surf was too low for board-riding and the weather unsuitable for sun-bathing, I spent time hunting for fossils in tall cliffs of sandstone and shale from the Triassic period of geological time, some 248 million years ago. Cracking and levering out large slabs sometimes exposed beautiful, large, fossil leaves. Some of these were fossil ferns and horsetails, and others were extinct fernlike seed plants, tantalizing relics of a very different ancient world. The leaves were not scattered throughout the shales, but instead confined to a few horizons. The thin layers crammed with fossil leaves also showed crumpling, skeletonization and other decay of the leaves, like a litter of partly decomposed leaves covering a soil. Roots of the fossil plants were preserved also, as irregular black streaks, branching down-ward at all angles through the rock beneath the fossilized leaf litters. These sedimentary rocks riddled with carbonized fossil roots were ancient soils!

They were not just fragments of an ancient soil, like the red nodules littering most of the Sydney landscape, but whole paleosols preserved intact within the rocks. I was elated to come to this stunning conclusion as a teenager, when I could find little written or spoken of paleosols. I later discovered that paleosols in sedimentary rocks are inferred in parts of James

Hutton's 1795 "Theory of the Earth' and charmingly called "dirt beds" in William Buckland's 1837 "Bridgewater Treatise", so have been known for as long as geology has been a formal science. Buried soils in the alluvium of the Danube River were recognized by Luigi Marsigli in a military engineering report to the Austro-Hungarian Emperor published in 1726. The idea that soils could have a fossil record was a personal epiphany that has guided my study and research ever since.

The idea of a fossil soil is a simple, but at the same time, a powerful idea because it suddenly made sense of a variety of formerly disparate observations. It is a rare moment in the life of a scientist when a simple idea grows into a comprehensive hypothesis, and the paleosol idea really had legs. The gray rocks were riddled with carbonized root traces pass raggedly down into reddened clayey rocks, just like the gray surface and red subsurface horizons of soils formed under forests. The layers associated with fossil root traces also have a complex system of cracks. None of the cracks are still open. They were crushed long ago by the tremendous weight of overlying rocks. Former cracks are now marked by linings of dark clay washed down to coat their walls, or by stains of rusted iron minerals. The complexity of overlapping and cross-cutting cracks was evidence that they opened and closed many times over, just as in soil wetted by rain and dried by the sun. Roots, changing textural horizons, and soil cracks familiar from our garden were preserved in rocks from millions of years ago.

Animal fossils are surprisingly rare in the sea cliffs, but the paleosols betrayed other evidence of soil animals familiar to me from childhood dirt-grubbing. Sinuous tubular burrows filled with clayey pellets resemble holes made by earthworms, littered with castings the creature passed, and then partially collapsed (Fig. 1.4). Other sand-filled vertical burrows are like escape passages of cicada nymphs emerging after many years of feeding on roots underground. Near Sydney still, brown cicada nymphs crawl out of the ground and crack open along their hunched backs. The brilliant green adult (*Cyclochila*) emerges from the cracked, brown carapace, spreads and dries its crumpled wings, then flies off to drone through the summer. It is remarkable to find hints of similar emergence from soil in rocks hundreds of millions of years old.

By the time I had the university training and means to make microscopic observations and chemical analyses I discovered further evidence of exactly what kinds of soils they were. The paleosols have clay-depleted horizons like those formed in soils by acidic leaching, spicular silica like that of accumulated plant opal, and a host of other soil features. When small slabs of paleosol-rock are ground thin enough to be transparent and examined under

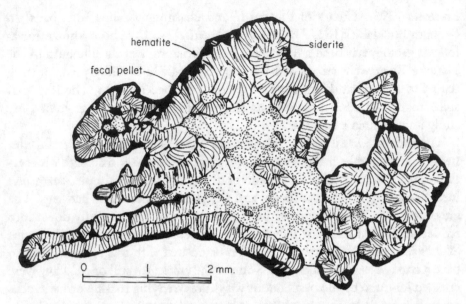

Fig. 1.4 Cross section of an earthworm burrow, filled with fecal pellets, some 248 million years old, from a fossil soil north of Sydney, Australia. The margin of the burrow has collapsed irregularly, but not before encrustation with the iron-carbonate mineral, siderite, which has partly rusted to the iron oxide mineral, hematite (reprinted from Australian Journal of Earth Sciences, 1976, with permission of Taylor and Francis)

a microscope, the system of cracks and alteration can be seen in fine detail. When analyzed for chemical and mineral composition, the distribution of elements and minerals through different levels of the paleosols is similar to that of particular kinds of soil. With a little armchair travel in the university library I discovered that these paleosols were leached and silica-rich like the soils of England and Massachusetts, not calcium-rich like desert soils, nor aluminum-rich like tropical soils. Here in the sea cliffs of Sydney were fossil soils much older and very different from those of the Miocene jungles of 16 million years ago. These soils of the Triassic some 248 million years old were more like those of cool temperate, humid woodlands and coastal heaths. Previously unseen landscapes of the past had now come into sharp focus (Fig. 1.5). My childhood questions were finding answers.

These answers were an odd new geological research direction which did not quite fit with what I was being taught at university. My undergraduate days were at the height of a boom in the global exploration for oil, which was guided by a sedimentary view of the rocks I was studying. Many of my fellow students became well-paid oil geologists, with expensive hobbies, and busy travel schedules to remote oil fields. They had no time, nor inclination to

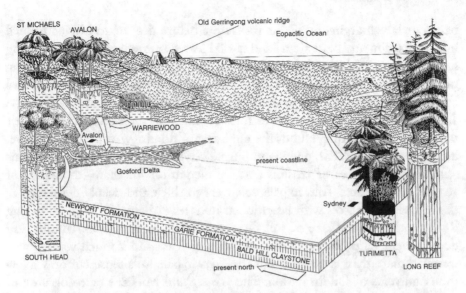

Fig. 1.5 Reconstruction of Triassic (248 million year old) soils and plants now exposed in the sea cliffs north of Sydney, Australia. The Bald Hill Claystone and Newport and Garie Formations are formally named units of rock whose environments of deposition are reconstructed in this diagram. The names in capitals are a variety of paleosol types, each representing a distinct ancient ecosystem (reprinted from Australian Journal of Earth Sciences, 1977, with permission of Taylor and Francis)

consider soils and agriculture. Glimpses of ancient landscapes and their soils were unexpected in the vast lowland basins envisaged for the accumulation of coal and oil. Such sedimentary rocks are thought to form by deposition of small particles, rock fragments and minerals, laid by currents of wind, waves and water flow. By that view, the sediments around Sydney were thought to have been deposited in large lagoons and rivers. Ancient sandy beds of rivers are well exposed and their sinuous courses can be mapped in the coastal cliffs near Sydney. Thin shale beds like those formed in lakes and lagoons confirm this interpretation. Such sedimentary rocks form by the flow and flooding of rivers, and by settling of mud in lakes. By the sedimentary view, all is flux, motion and accumulation, not slow alteration and growth in place of plants and soils.

The presence of paleosols indicated, however, that a concept of sedimentary accumulation should be modified to include periods of soil formation, when sediment was not being deposited. The sequence of rocks in the sea cliffs records the efforts of vegetation to bind and use sediment in the intervals between catastrophic burial or erosion by floods. Some of the buried soils are little altered from sediments, and their original sedimentary layering is barely disrupted by root traces. These buried soils indicate colonization of river banks or lake margins by plants for a time before destruction by additional influx of sediment. Bent stems and curled leaves in the base of the overlying sandstones are evidence that small shrubs and saplings succumbed to rushing turbid waters. Other buried soils showed leached zones, reddened zones and complex crack systems that entirely obliterated the original bedding. In these ancient soils, the saplings grew to forests, churning the sediment with their roots, and withstanding all but the mightiest of floods. Within these massive, hackly sequences of red, green and purple claystones, old-growth ecosystems withstood the forces of erosion and sedimentation for many thousands of years. None of these Triassic paleosols were so thick and deeply developed as the Miocene paleosols with lateritic red nodules at Frenchs Forest and Long Reef. Studies of the rate of soil formation indicate that thick lateritic soil development takes hundreds of thousands of years and a much warmer and wetter climate than at present, but the Triassic paleosols represent only a few thousand years of soil formation. The Triassic and Miocene paleosols around Sydney represent triumphs of the living world against the forces of erosion and sedimentation.

Derek Ager, a well-known British geologist, coined a memorable metaphor for the sedimentary record, which "like the life of a soldier, consists of long periods of boredom punctuated by short periods of terror." The record of sedimentation, volcanic eruptions and meteorite impacts reveal battles and revolutions of the past. In contrast, paleosols represent the cumulative effect of ordinary life. Soil formation is as slow as watching the grass grow, but it is relentless and fundamental. The opposed forces of soil building and destruction are themes to which we will return on a variety of scales from the daily work of lumberjacks to the rare devastation of asteroid impact.

With my eyes now open to the fossil record of soils, I thought again about fossils, and particularly about fossils in fossil soils. Sometimes plant and animal fossils are found within the soils that supported them, as in the plant-bearing paleosols near Sydney. In these cases, information from fossils and paleosols combined gives a vivid impression of past land environments. Fossils of land creatures are also found in deposits of lakes and deltas, removed from the soils that nurtured them. These transported fossils provide a less grounded view of past environments. There are also paleosols containing no identifiable fossils, such as the Miocene lateritic paleosols around Sydney. Why was that? I had collected many skulls and other bones from dusty and salty soils while visiting boarding school friends on sheep stations of inland New South Wales, but how many bones had I seen in many years of teenage bushwalking in the forests near Sydney and in southwestern Tasmania? None. Dry alkaline rangeland soils do not destroy bones and teeth. In contrast, forest soils are rich in acids which etch away the calcium phosphate of fossil bone and shell. Forest soils also are too well drained to prevent decay of leaves and soft tissues of animals. The fossil record of well drained rain forest ecosystems is poor because rain forest soils are too acidic to preserve bone and too oxidizing to preserve leaves. Lateritic paleosols may be all that is left of these stately and diverse ecosystems. With the direct fossil record of plants and animals so incomplete, paleosols may be the only remaining evidence of some past worlds. Gaps in the fossil record could be filled with information from paleosols.

Compared with fossils, paleosols present a different kind of evidence for ancient life. Paleosols are a record of environmental alteration by ecosystems, not a record of individual extinct creatures. The overall form of a paleosol shows the net effect of an entire ecosystem on its immediate environment. It reflects work done on the landscape, as does the fossil burrow of an earthworm. The fossilized body of an earthworm is one kind of fossil, a body fossil, but the burrow remaining from the earthworm's activity is another a kind of fossil, called a trace fossil. Paleosols are, in a sense, trace fossils of ecosystems. They are records of ecosystem processes, such as long-term plant nutrition and land stabilization. They reflect the general impact of life at its most fundamental level. Thus fossil soils, and the fossils they contain, offer different but complementary perspectives on life and landscapes of the past. Paleosols supply new vistas of the evolution of life and soils in the distant geological past.

I came to see that soil is more than a foil for childish energy, but a profoundly important part of our world. Life and soil support each other in intriguing ways. Living creatures, with their roots, jaws, and other means of acquiring nutrients, do much to determine the nature of soil. Conversely, the nutrient resources and drainage of soil determine what kinds of life can thrive in a particular place. Soils are a fundamental part of terrestrial ecosystems, and form a complex interface between the living and inanimate worlds. Soils contain raw materials such as minerals, air, and water. They also contain products of weathering, which include not only weathered clays and nodules, but the bodies of organisms as well. Soil includes billions of bacteria, millions of nematodes and a few plants in just about every cubic inch. The soil's diverse microbes and internal surfaces absorb and destroy poison and other noxious products of decay. Soil purifies water that flows through it and regulates the composition of the atmosphere by setting a balance between photosynthesis and decay. Through soil, life has far-reaching effects on land, water, and air.

Soil is mother earth, worshipped of old as Demeter of the Greeks or Ceres of the Romans. We owe much to her, and to her daughter Proserpina, whose familiar myth begins with that carefree girl picking flowers on the grassy slopes of Eleusis (now Elefsina near Athens). Spring asphodels, chrysanthemum and bellflower still bloom there, overlooked by a limestone crag with twin caves like the eye-sockets of a giant skull half buried in the meadow. Suddenly Proserpina was seized by Pluto and carried off to the underworld though this cave. When the loss of her daughter became apparent, mother Ceres, the goddess of grain, plunged the world into a winter of sorrow and failed crops. Eventually Ceres discovered where Proserpina had been taken and negotiated her return, but not without cost. Because Proserpina had tasted the pomegranate in the underworld, she was a codependent, and condemned to return to the underworld annually with the cycle of the seasons (Fig. 1.6). Proserpina was emblematic of contrasting seasons, the innocent exuberance of spring versus the leaden skies of winter. Proserpina's seasonal rhythms of life have also been rhythms of soil. The spring flush of wildflowers and new grass, then bare-branched trees coming into clusters of small, light green leaves, is a frenzy of plant production after a long winter of plant dormancy. The soil is then coursing with warm spring water, which mobilizes plant nutrients such as calcium and potassium from mineral grains, such as feldspar, leaving behind a residue of clay-skins within the soil cracks. By fall, the long summer of plant activity comes to an end, as leaves blaze with color, drop and decay. Bacteria and fungi turn the leaves into intricate skeletons of veins, then into fine dark humus. Earthworms and termites drag leaves and twigs down into their holes for further consumption. This rhythm of spring plant production and fall plant consumption in the northern hemisphere is so extensive that it alters the composition of the atmosphere. Concentrations of carbon dioxide in air rise by as much as five parts per million by volume during the fall and winter of decay and consumers. Carbon dioxide falls by a comparable amount during the spring and summer of photosynthesis and producers. I have called this seasonal breathing of the world's ecosystems, the Proserpina Principle, the simple idea that plants cool the planet, but animals warm it.

Evidence for comparable rhythms on longer time scales comes from paleosols. Over thousands of years, soil fertility declines, or peat or salts accumulate, so that plants can no longer produce leaves at the same rate and quality as before. The animals that eat the leaves still find much to eat until they overgraze and trample their forage. Eventually wind and rain pluck soil exposed between the thinning plants into dustbowl storms that are catastrophic for plants and animals alike. As the dust settles however, it recharges

Fig. 1.6 The ancient Roman queen and king of the underworld, Proserpina and Pluto, from a votive stele at Locroi Epizephuroi, on the toe of Italy (480–450 BC). She holds a cock and the spring crop of barley. He holds a wine bowl and a branch of autumn pomegranates

the new soil with minerals that fuel more vigorous plant growth than before, at least until animal stocking densities recover from the hardships of dust-bowl years. Do millennial cycles of plant, then animal dominance, affect the atmosphere and climate as well?

On even longer time scales of millions of years, the Proserpina Principle could have worked though alternating evolutionary innovations. The balance may have tipped toward animals as they evolved jaws, strong teeth and hard hooves, and then tipped toward plants as they evolved tough bark, strong wood and insecticides. Does the coevolutionary interplay of plants and animals, with their divergent agendas, explain long term evolution of life and soil on land? There were also global disasters for life from volcanic eruptions and meteorite impacts when the oxidation of life and soil created alarming and short-lived greenhouse warming. Did atmospheric rhythms cued to cycles of life in soils allow ecosystems to roll with these brutal punches?

How did this complex machinery evolve over geological time? Were major advances in the evolution of terrestrial ecosystems, such as the advent of forests, a biological imposition on soils? Were they facilitated by soil development? Or were soils and their communities themselves engines of global change? In the very beginning, was there soil without life or life without soil? Or did they evolve together, or from one another? This book is dedicated to the proposition that paleosols provide an historical perspective on how the world has worked over the long sweep of geological history. Soils and paleosols offer a different perspective, a worm's eye view, of global change and life's origins.

RELICS OF THE PAST

Stones of Sydney are stained red with age,

Blackened by fires that in summer heat rage

Through resinous gums, and thorny shrubs.

On broad rock benches are long curved grubs

Of stones in rows, carved grooves, and forms

Of Dreamtime beings in Koori norms,

Made by men of old, before white sails

Brought to these shores a world of nails.

Older than nails and tools of stone

Is China clay bleached white as bone,

And red clayey nodules deep in soil,

Relics of rain forest's ancient toil.

2

Soil as a Many Splendored Thing

Soils vary not only within their profiles, but geographically. Different kinds of soils represent different local conditions of climate, drainage and other factors, which can be inferred from paleosols.

For many people soil is dirt, a substance beneath consideration, a nuisance to be eradicated from our lives. But dirt is just soil out of place, like a weed in the garden. There is monotony to house cleaning and ordinariness to the soils in our immediate neighborhood. It is not until you dig, look closely, or travel widely, that the spectacular variety of soils becomes apparent. Soils vary greatly from the surface to the bottom of a single hole in the ground. Vertical variation in soils reflects their unique position at the nexus of ecosystems and air on the one hand, versus rock and sediment on the other hand. The variation of soil horizons reflects a gradient of alteration down from the land surface. The late Francis Hole of the University of Wisconsin once regaled us at meetings with the following song to the tune of "On top of old Smokey" (poor choice of tune, but great lyrics).

A rainbow of soil is under our feet,

Red as a barn and black as a peat,

It's yellow as lemon and white as the snow

Bluish gray, so many colors below

© The Author(s), under exclusive license to Springer Nature
Switzerland AG 2022
G. J. Retallack, *Soil Grown Tall*,
https://doi.org/10.1007/978-3-030-88739-1_2

Even more striking variation in soils can be seen in different parts of the world. Landscapes are a mosaic of different kinds of soils, each soil uniquely suited to local conditions. The great diversity of soils reflects in part a long evolutionary history of the ecosystems that they support, as outlined in future chapters. For now however, consider some fundamental features of soil horizons and soilscapes as if they were tourist destinations to explore and enjoy.

The surface layer of soils, or the A horizon in the international shorthand of soil science, is the one familiar to gardeners, farmers, and grubby little boys. It is often a gray mix of loose organic debris and mineral grains. This is the layer richest in living things, including plant roots and teeming multitudes of small animals and microbes. This is the fertile layer that gives us carrots, cabbages, and corn. For gardeners and farmers this is the material that really matters, and its moisture, texture, fertility, and organisms are carefully maintained. For soil scientists and geologists however, there is more of interest deeper in the soil.

Beneath the A horizon and commonly within reach of roots, many soils have more clayey and rusty clods, called a B horizon. Unlike the loose, gray A horizon, the B horizon commonly is stiff with red and brown clay, which forms thin coatings and angular, potato-sized clods. The clayey B horizon is designated Bt in the international shorthand of soil science. The "t" is short for *ton*, which is German for clay. Soils of the Sydney region in Australia are of this type, formed under eucalypt woodlands of this humid coast. My first exposure to these Bt horizons came as a boy. Inspired by a children's book showing dinosaur skeletons in layers deep below a city, my friends and I began to dig a hole under the house. We knew that, tempting as it was, digging through from Sydney to London of the illustration was probably beyond our capacity, but surely there must be a dinosaur skeleton within a few hundred feet of our house. There were many skeletons in the illustration. Three of us managed to remove the loose topsoil or A horizon, in a hole about 2 ft. deep by intermittent activity over a period of several days. The clayey layer beneath proved extraordinarily tough, especially for six-year olds. The soil clods there were large, sticky and hard. After what seemed like ages, we gave up, and urinated into the hole in disgust.

Beneath the B horizon of a soil is weathered rock or sediment, the C horizon in soil science shorthand. It is usually less than 6 ft. down, the standard depth for grave-diggers. Here we can rest in peace beyond the roots of trees and burrows of most animals. The C horizon in the Sydney area is bleached white, with orange mottles and red nodules remaining from a much earlier period of unusually deep weathering, but it remains in most

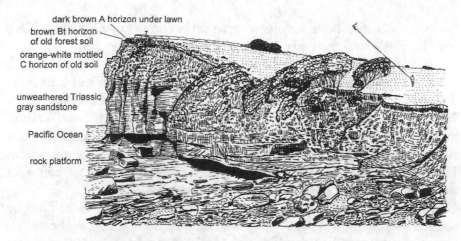

dark brown A horizon under lawn
brown Bt horizon
of old forest soil
orange-white mottled
C horizon of old soil

unweathered Triassic
gray sandstone

Pacific Ocean

rock platform

Fig. 2.1 A thick soil from an ancient forest, now cleared as a city park, at the coastal cliff called the Skillion, in Terrigal, New South Wales. Here the deeply weathered C horizon is orange and white and extends 18 ft. below the topsoil to fresh, gray-green rock. The soil consists of a dark gray, surface (A) horizon over a brown, clayey, subsurface (Bt) horizon. The man flying a kite gives an idea of scale

places as a C horizon to the current soil as well (Fig. 2.1). The discolored C horizon is called a saprolite, but below that can be another part of the C horizon called saprock, that looks fresh but is riddled with cracks. Solid bedrock is even further down, reached by deep road cuts, major excavations, and sea cliffs. Around Sydney, bedrock is hard, white sandstone, with local layers of dark gray shale. Bedrock is hard enough to ring to the crack of a geological hammer, and needs heavy equipment to move. During excavation it needs to be drilled and blasted with dynamite. By contrast, saprolite and saprock, while looking like rock in some ways, can be moved by bulldozers without recourse to blasting. Civil engineers and geologists still regard saprolite as part of the soil profile, though to farmers and gardeners this is not soil of consequence. Thus, soil horizons have both engineering and agricultural implications.

Not all soils have distinct A, B, and C horizons. These are the typical horizons of forested soils in humid climates. Some soils have more horizons and some have fewer, depending on their climate, vegetation, and other factors. For example, swamp soils have peat layers, and are called Histosols from the Greek *histos* for tissue, in this case plant tissue, which can be seen under the microscope. Permafrost soils with extensive ground ice are called Gelisols, from Latin *gelare*, to freeze. Weakly developed soils without marked horizon differentiation are called Inceptisols, from the Latin *inceptum*, beginning. Soil can be quite varied both within a single profile and between different kinds of soils, which have their own personalities and names (Fig. 2.2).

Much variation in soils can be seen if one travels with an eye for earth, as I did as a teenager trying to discover the world. My friends had heard that there was work to be had in Wee Waa, on the plains of inland New South Wales, Australia. It was supposed to be simple work "chipping cotton" (that is, weeding a crop of cotton), with good pay. We got the jobs remarkably easily and were feeling quite blessed until we made three disillusioning discoveries. The pay was miserable. The work was arduous in intense summer heat. Our work mates included genuine psychotics and criminals.

Nor were we warned about those unforgettable black clay soils. The clay cracked deeply and was baked at the surface into clods as stiff as cheese. We filed and sharpened our hoes for ease of slicing the hard earth and fibrous weeds. The clay presented yet other problems after dowsing by thunderstorms that circled endlessly over the parched plains. Once wet, the clay was so sticky that we could not walk more than ten steps from a row of cotton plants without taking off boots caked with clay to the size of footballs. At times after heavy rains, not even the bus could move on roads prepared from such soils to take us back to the working men's barracks.

These clayey soils are called Vertisols, full of clays that have the unusual properties of stickiness and swelling when wet. The name is derived from Latin *verto* (I turn) in reference to their instability through wetting and drying cycles. Houses and roads built on such soils tend to buckle and deform as the clay expands and contracts with wetting and drying. I have since swerved through miles of wet bush roads on Vertisols in Kenya and seen the pervasive systems of fractures in deep pits cut into such soil in the coastal plain of Texas. India also has thick, dark, swelling-clay soils that cap its high basaltic plateaus. Vertisols form in areas of swelling clays where weathering is significant, but not profound, usually under grassland and grassy woodland in climates showing a marked dry season.

Our experiences in Wee Waa were an adventure that we did not care to repeat, so we took a much longer drive the next summer and were amazed to see the vast extent in the Australian outback of desert soils, or Aridisols (Color Photos 2.1, 2.2). This soil name is also from Latin, *aridus*, for dry. We found work 400 ft. below the ground in a nickel mine at Kambalda, near Kalgoorlie in Western Australia, but there was time during weekends to take in some sun and soils. Exploring a salt pan by car one weekend, I was soon bogged up to the axles, 15 miles from the closest help. My big mistake was stopping. As long as the car was rolling, the salt crust at the surface was strong enough to keep the wheels above the underlying soft clays. Getting the car

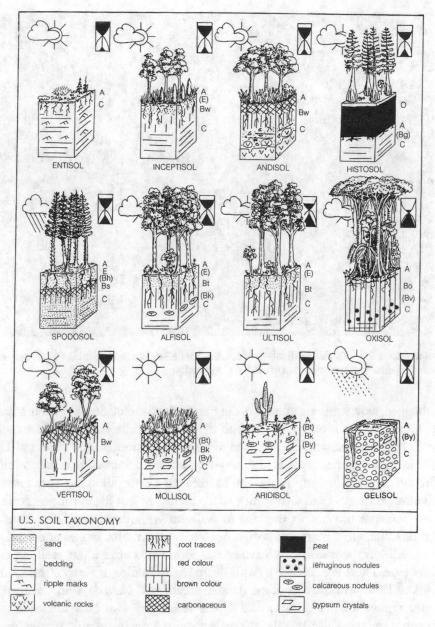

ENTISOL

INCEPTISOL

ANDISOL

HISTOSOL

SPODOSOL

ALFISOL

ULTISOL

OXISOL

VERTISOL

MOLLISOL

ARIDISOL

GELISOL

U.S. SOIL TAXONOMY

sand

bedding

ripple marks

volcanic rocks

root traces

red colour

brown colour

carbonaceous

peat

ferruginous nodules

calcareous nodules

gypsum crystals

Fig. 2.2 Cartoons of different kinds of soil recognized by the Soil Conservation Service of the U.S. Department of Agriculture. The sketches of vegetation are at a different scale than the soil profiles, which are in blocks 5 ft. high. Different climatic regimes are suggested by sun, cloud and rain symbols. Time for soil formation, suggested by hourglass symbols, ranges from only a few years in Entisols to hundreds of thousands of years in Oxisols. Letters beside the soil columns are a shorthand system of labels for different kinds of soil horizons, and those in parentheses are common but not essential to the definition of that soil type (reprinted from Retallack 2019, Soils of the Past, with permission of J Wiley and Sons)

Color Photo 2.1   Calcic Aridisol, under red mallee (*Eucalyptus socialis*) woodland, near Damara Station, New South Wales, Australia

unbogged took some time, much digging, a measure of despair, then a long walk to help, but it taught me a good deal about Aridisols. These desert soils are full of salts that would be flushed out of soils in more humid climates. The salts include gypsum crystals and powdery white calcium carbonate nodules. The salts crunch underfoot and are harmful to plants, allowing only growth of scattered desert shrubs. Without shade plants, or leaf litter, the soil surface itself becomes too hot in the sun to walk on without shoes. There is little turf or soil organic matter to soften one's footfall on salts, stones and sand. A long walk on these soils can leave one as footsore as after a day walking city pavement. There are extensive Aridisols in the rift valley of Kenya, the barren coast of Peru, and intermontane deserts of Arizona, Nevada, and California. These regions of dry climate and sparse vegetation also lack sufficient soil moisture to etch soil minerals, to create a soil reserve of organic matter, or to dissolve soluble salts.

My first academic job in the United States, in Dekalb, Illinois, turned into a pilgrimage to one of the finest agricultural resources in the world. The till plain of northern Illinois supported tall grass prairie on fertile soils called Mollisols, but now is largely planted under corn. Although thick and dark like the Vertisols of Wee Waa, Mollisols are very different in structure and

Color Photo 2.2   Shallow-calcic Aridisol, under bluebush (*Maireana sedifolia*) shrubland, Mungo Lakes National Park, New South Wales, Australia

texture. Their clay is intimately admixed with organic matter, which stabilizes the clay within small, rounded clods usually a tenth of an inch or so in diameter. These little clods, or crumb peds as soil scientists call them, run through the fingers like tiny, soft pellets. The clods are not hard like rock or mineral grains, but a mixture of tiny mineral grains, clays, and organic matter, easily crushed between the fingers. This soft and spongy surface is the source of the soil name Mollisol from the Latin *mollis*, soft. Mollisols are the most remarkable and economically important soils: stable after rain, rich in organic matter, and exceptionally fertile. The distinctive crumb peds of Mollisols were created over thousands of years as fecal pellets of earthworms and as the matrix to a network of slender grass roots. Grassland soils form a broad climatic belt. They receive less rain than the deeply weathered forest soils of Ohio and Indiana, but more rain than desert shrublands of Wyoming and Colorado. The tall grass prairie region now is a patchwork of corn fields and other croplands. Other similar areas of high productivity grassland soils are found in the southern Russian Plain and the Argentine pampas around Buenos Aires. These regions remain the bread baskets of the world.

From Dekalb I moved to another job east into the rolling forested hills around Indiana University in Bloomington, southern Indiana, and another soil type, called an Alfisol. This name was formed as a compound from the alumina (Al) and iron (Fe) that dominate the chemical composition of their distinctive red-stained clays. The oak (*Quercus*) and hickory (*Carya*) woodlands of the Indiana have leaves that are soft and bright green, unlike the leathery, olive-colored leaves of Australian eucalypts. Australian eucalypts are evergreen, but oak and hickory trees lose every leaf in a blaze of fall color. It was amazing to me, accustomed to year-round plant and animal activity in Australia, to see plant production shut down entirely for the Indiana winter. How could trees afford to throw away the great piles of leaves collected from Bloomington's streets each fall? There are equally harsh winters in the southeastern Australian Alps and in southwestern Tasmania, yet snow gums (*Eucalyptus pauciflora*) and southern beech (*Nothofagus cunninghami)* there retain their leaves above and under the snow. The leaves of snow gums persist through summers as hot as those of southern Indiana. Southern beech (*Nothofagus menziesii*) forests of New Zealand are evergreen, whereas South American beech includes both evergreen (*Nothofagus betuloides*) and deciduous species (*Nothofagus antarctica*), and both places also have snowy winters. Climatic differences do not appear to explain why some forests shed their leaves while others do not. Soils may be a part of the explanation.

Soils of southern Indiana have a gray organic surface (A) horizon over a reddish brown clayey subsurface (Bt) horizon, outwardly like those of southeastern Australia. Many forest soils have these A, B and C horizons, with roots extending down to about 3 ft. The main difference between Indiana and Sydney soils is one of fertility. The Indiana soils are very fertile, as can be seen from their generally alkaline pH, and abundant plant nutrients. Calcium for example, is needed for cell walls, and magnesium for chlorophyll. Potassium and sodium are important cell electrolytes. Phosphorus is essential for nucleic acids and enzymes. The high chemical fertility of Indiana soils is apparent from their clays, particularly the clay known as smectite, and the persistence of minerals such as calcite. The soils in turn derive their fertility from glacial silt blown by Ice Age winds, from nearby sandstones full of feldspars, and from hills of limestones and shales. The Indiana Alfisols are nutrient-rich forest soils. They can be contrasted with Sydney area Ultisols, which have similar A, B and C horizons, but an acidic pH and nutrient-poor minerals, such as kaolinite and quartz. There is no calcium, magnesium, potassium, sodium or phosphorus in kaolinite or quartz, which are mainly aluminum, silicon and oxygen. This chemical composition is the result of a long period of weathering implied by the roots of the term Ultisol, from the Latin *ultimus*, last. Trees acquire nutrients slowly in many Australian Ultisols and do not shed such hard-won biomass, but in Indiana there are such plentiful mineral resources in the soil that the difficulties of winter are bypassed by dormancy, and the shedding of leaves that are soft and easily consumed by earthworms. The oak forests of the American midwest evolved in Alfisols of plentiful nutrients, whereas eucalypts and southern beech forests of Australia evolved in impoverished soils, and have remained evergreen even when colonizing fertile, young, montane soils of New Zealand and South America. There is a strong seasonal rhythm of summer photosynthesis and winter respiration in Indiana forests, but not in Australia or New Zealand. South American forests have both patterns with both evergreen and deciduous species on fertile soils. Climate is not the sole determinant of vegetation; soils play a role as well. Soil, life, and air are intimately interconnected.

Soils of the United States are as varied as its vegetation and climate, with little relation to bedrock or sediments. Ultisols like those of Sydney, Australia, formed under the tall forests, warm climate, and stable landscapes of the southeastern U.S., from Virginia south into the coastal plains of the Carolinas and Georgia. These support a mix of deciduous plants such as oak (*Quercus*) and basswood (*Tilia*), as well as evergreens such as pine (*Pinus*). Further north in the Catskill and Adirondack Mountains of New York state, a cool rainy climate and quartz-rich local rocks have produced another kind of soil with a distinctive subsurface zone rich in iron, humus, and quartz. Such clay-poor, acidic, gray soils are called Spodosols, from the Greek *spodos* for wood ashes. They support evergreen forests of spruce (*Picea*) and pine (*Pinus*) that tolerate soil nutrient shortages. Alfisol, Ultisol, and Spodosol are technical terms, but they signify important soil differences that reflect natural ecosystems and potential use by humans. Alfisols for example are suitable for a variety of agricultural uses, such as pasture, market gardens and orchards, but Ultisols and Spodosols are best left as native forests or for watershed protection. Wetlands such as salt marshes have grey, little weathered, silts (Entisols) more like rooted sediments than other soils (Color Photo 2.3). Nationwide mapping of such soil types county by county on a scale suitable for landowners to find the kind of soil on their property has made the U.S. Natural Resource Conservation Service (formerly Soil Conservation Service), and its soil classification scheme, an invaluable component of national prosperity.

National mapping and classification of soils has long established the overall pattern of soil, climate, and vegetation in the United States (Fig. 2.3): forested Ultisols in the Pacific Northwest, desert Aridisols in the intermontane west, grassland Mollisols in the Great Plains, then nutrient-rich Alfisols under deciduous forests of the midwest, and nutrient-poor forested Ultisols and

Color Photo 2.3   Bohicket soil, Gleyed Entisol, with subsurface pyritized shells, under salt marsh (*Spartina alterniflora*), intertidal zone of Bass Creek, Kiawah Island, South Carolina

Spodosols to the east. This pattern follows an east–west climatic moisture gradient of high rainfall in western Oregon and Washington, but an extended rain-shadow east of the Cascade and Rocky Mountains. From the dusty dry western parts of the Great Plains near Denver and Lubbock, rainfall increases to the east into the rolling farmlands of Pennsylvania and the misty forests of the Great Smoky Mountains. Intermontane Aridisols of the western rangelands have abundant nutrients but insufficient natural soil moisture for crops. Along both the northwestern coast and the eastern seaboard and Appalachians, high precipitation has strongly depleted reserves of mineral nutrients from Ultisol and Spodosol soils. In the Great Plains and Mississippi drainage, however, moderate rainfall and high fertility of glacial windblown silt conspire to create one of the most productive agricultural regions in the world on Alfisols and Mollisols. The pattern of national prosperity comes down to earth.

Another broad pattern of soil types follows the global north–south gradient of temperatures with latitude. Differences between soils from the equator to the poles can be seen on a smaller scale on tropical mountains. In few places is this more marked than in Peru, where snow-capped peaks of the Andes

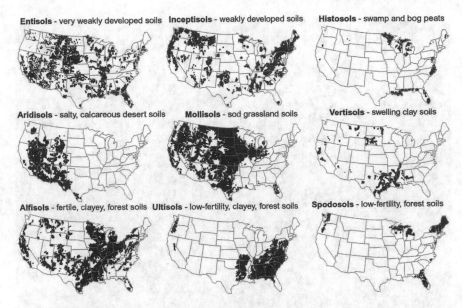

Fig. 2.3 Maps of the distribution of 9 of the 12 soil orders in the conterminous United States. Other soil orders are too limited in area to represent at this scale. The other three orders, Gelisols, Andisols, and Oxisols, are of very limited distribution in this area (reprinted courtesy of U.S. Department of Agriculture, National Resource Conservation Service)

fall away into the jungles of Amazonia. Like many young Australian mountaineers, I was lured to Peru by the promise of mountains, ice formations and archaeological ruins on a scale much larger than found in Australia or New Zealand. Intimidated by the size of these mountains, we hired mules and a mule driver to carry a month's supply of food and mountaineering equipment from the railway station at Santa Teresa, over rough mountain trails to a base camp just over Yanama Pass, at 15,680 ft. (4780 m) in the Cordillera Vilcabamba.

The tropical jungles of the valley of the Rio Vilcabamba near Santa Teresa grow in thick red soils on the margins of the vast Amazon River basin. Here weathering is promoted by copious rain, high temperatures, abundant termites and ants, and the mighty multistoried tropical rain forest. The soils are deeply weathered of nutrients, have alumina-rich clays, and are full of red, clayey, fecal and oral pellets of termites. These are deeply weathered tropical soils, or Oxisols (Color Photo 2.4), so named from the French oxide, in reference to abundant oxides of aluminum and iron, and in acknowledgment of the fine research of French soil scientists on such soils of tropical Africa and South America.

Color Photo 2.4   Kahuna soil, Oxisol, under cut koa (*Acacia koa*) forest, on Maunaloa Hill, Molokai, Hawai'i

An ecological paradox of Amazonia is the luxuriance of the forest and exuberant diversity of animals in a soil that is nearly exhausted of mineral nutrients. This impressive community is sustained by efficient recycling, for example, by fungi, which break down dead organisms into their constituent elements. These vital forest litters and the uppermost soil horizons are the reason that agriculture in this region has traditionally been practiced on a swidden system. By this method, small fields are slashed and burned, then planted with crops until the nutritious organic surface is depleted to nearly sterile, hard-setting, red clays. After only a few years, the fields are abandoned for new clearances. The swidden agricultural system has worked for thousands of years, but permanent ranches on these infertile soils have failed as the soil develops thick, hard-setting crusts, or is eroded into unstable badlands. Oxisols are best left forested, or planted with a mix of tropical tree crops, such as coffee, cocoa, and bananas. These cultivated plants lined the path into the spine of the Cordillera Vilcabamba as we trudged behind our flatulent and stubborn mules.

At 8000–10,000 ft. (2400–3000 m) along our Andean mountain trail we found welcome cooler temperatures and a set of crops and soils more like those of temperate regions. Every spare inch of level ground, and hillsides excavated into artificial terraces, was planted with corn and potatoes

Fig. 2.4   Salcantay Peak (17,293 ft. or 5271 m), girt by glaciers, rises above the clouds and Amazonian rain forest, in the Cordillera Vilcabamba, Peru. Here young glacial Gelisol soils are within view of ancient, deeply weathered, Oxisol soils of tropical, rain forest

in brown soils similar to the Alfisols of Indiana. This was sweet corn (*Zea mays*), not wheat or other cereals sometimes called "corn" in Europe and the Middle East. Originally these soils probably supported mountain fuchsia forest. Native fuchsia forest still grows in patches beyond Quechua villages, which at this altitude are constructed of stone rather than wood.

Above 12,000 ft. (3600 m), the fuchsia forest gave way to grassy alpine meadows with llamas, alpacas, sheep, and cattle (Fig. 2.4). At our base camp at 15,000 ft. (4570 m), we were surprised to see stunted potato plants, and after little digging, found potatoes the size of peas. The rows of former cultivation, now grassed over, were clearly visible from higher vantage points. The potatoes were probably planted hundreds of years ago, but the climate now is too cold and has too short a growing season for useful potatoes at such an elevation. Like the massive Inca stonework of nearby Macchu Picchu, these furrows are remnants of ancient human occupation. The soils are thin and rich in organic matter, but with little warmth and water, little nutrition is passed on to plants. Nor is the soil at such high elevation host to such a diversity of microbes or such large roots as lower in the mountains. These alpine soils retain sedimentary layering from sheet wash of snow melt over alluvial fans. Such soils with little evidence of weathering other than plant rooting and disruption from ground ice are called Gelisols.

Higher still, life clings tenaciously to the rocky peaks. Below the zone of glaciers and ice-caps are lichens and red algal stains. I was amazed to find on one mountain top, pockets of humus-rich soil sustaining small native asters (*Chuquiraga*) at elevations of 17,000 ft. (5200 m). It seemed almost as out of place as two overdressed mountaineers with ropes, carabiners, and chocolate.

The variety of soils and plants encountered while climbing tropical mountains reflects changing environmental conditions with altitude, or ecological zonation. This should not be confused with ecological succession, which is the return of life as time passes after disturbance. Ecological succession brings back the jungle to swidden fields of Amazonia. Bare ground is at first colonized by algae and lichens, then herbaceous plants such as grasses that prepare the soil with humus and clay for the germinating seeds of saplings and trees. In ecological succession, unlike zonation, conditions change with time as a sunny open field is converted to the cool shade of forest. Ecological succession takes only a few years in a swidden plot of Amazonia. On much longer time scales of the evolution of the Earth and its ecosystems, one can imagine a geological succession that began with algae and lichens, later enriched with herbaceous plants and forest. The algal-lichen communities of mountain tops and disturbed ground can be viewed as a legacy of this long geological history. Soils, like plants and animals, have diversified to include ever more sophisticated types, as we shall see in later chapters.

Some of this geological history can be inferred from soils as they are today, as is apparent from foregoing comparisons of American and Australian Alfisol and Ultisol forest soils. Each soil type has agricultural and political limits, as well as a heritage of natural ecosystems. Paleosols are thus records of the geological succession of life and ecosystems on Earth. The following chapters explore fossil soils as evidence for cycles of global change in the geological history of soils. We begin with familiar soils of classical civilizations, then proceed back in time to unfamiliar soils of the distant geological past, and the origin of life itself.

SOIL SOURCE

Soil seems simple, but is more than dirt,

Swept or washed from a favorite shirt.

There's more than just one kind of soil:

Red, clayey soils of rain forest tall,

Deep brown loam on grassy plains,

Thin soil dust where rare are rains,

And thick, peat-moss on alpine fells.

The soil factory's dank, dark wells

And chimneys, roots and worms give,

From rock and sand, so millions live.

Soil is the stuff from which we came,

And until our end, waits to reclaim.

3

Civilization Built from Soil

Soils have been modified by civilizations, usually irreversibly, but remain the basis of human prosperity. Buried and deflated soils allow assessment of soil resources.

Our environment today is "of the people, by the people, and for the people", in a way that Abraham Lincoln could not have suspected in his well-known aphorism for democratic governance. We are now everywhere, changing everything, not only the buildings and fields around us, but also the air we breathe and the soil beneath our feet. The days when log cabins were hewn from the woodlands of Illinois seem such a long time ago amid the shopping malls and freeways today. Roads, dams, and parking lots are new kinds of land surfaces supporting only one kind of organism and its machines. New breeds of corn and grass, and application of superphosphate fertilizers, are staving off diminished soil fertility and the loss of soil organic matter oxidized by tilling and harvest. Nevertheless, soil humus has been absconded among the concrete pillars of New York and Hong Kong, as well as in the tailored agricultural landscapes of Illinois and England. At the same time as we are diminishing soil's capacity to absorb carbon dioxide from air by tilling and paving, we also are burning tropical forests and fossil fuels to add further to the atmospheric greenhouse and inexorable global warming.

Despite these environmental problems, technology thrives as we step onto the Moon, send missions to Mars, and put computers to work for millions. A correlation between major flowerings of civilization and warm temperatures is striking when one considers indications of past temperatures from

G. J. Retallack, *Soil Grown Tall*, https://doi.org/10.1007/978-3-030-88739-1_3

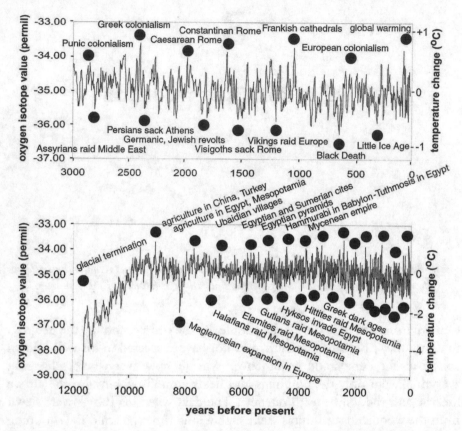

Fig. 3.1 Greenland ice core records of oxygen isotopes reveal fluctuation of temperature over the past ten thousand years. The isotopic values are relative to an arbitrary standard of a fossil shell from South Carolina, and the temperatures are inferred from laboratory experiments. Warm periods correlate with major flowerings of civilization, whereas cold periods correspond with disasters for human populations

gases trapped in the annual layers of Greenland ice cores (Fig. 3.1). Egyptian pyramids and Medieval cathedrals were built at times of global warmth and high carbon dioxide. In contrast, Dark Ages, the Black Death and the Viking raids all coincided with times of cold temperatures and low carbon dioxide. Was this because warm temperatures allowed human creativity, or because the pressure of human creativity creates warm temperatures? Bill Ruddiman of the University of Virginia blames human population growth for the warm spikes, and also for Holocene temperature maintenance for a lot longer than followed all other Ice Age deglaciations. The reason great things were done during warm spikes by this view, is not because the climate was pleasant, but because a lot of people unleashed unusual creativity and released enormous amounts of greenhouse gases. Before the industrial revolution the chief source of carbon dioxide was wood burning and agriculture. Land clearance and plowing both released soil carbon to the atmosphere. Plagues and wars curbed not only populations, but also reduced greenhouse emissions, with chilling effect. Some of these catastrophes were driven by crop failures and droughts which particularly afflict Mediterranean summer-dry climates during times of high temperatures. Human population growth thus oscillated within technological limits of the time to maximize carrying capacity, and stabilized climate against the long-term natural cycle that would otherwise push the planet towards the next glaciation.

A more nuanced view comes from three case histories of civilization versus soil, the vital link between civilized land use and the atmosphere. The long term rise and fall of civilizations can be linked to changes in soil during five millenia of soil-human interaction in Greece and England, but first consider the beginnings of human modification of pristine soils of New Zealand.

The Southern Alps of New Zealand present vistas of primeval wilderness as free of human modification as I have ever seen, but even in wilderness, signs of humans are difficult to escape. In remote corners of these mountains is a small plant called sheep's burr (*Acaena ovina*). Its fruit is a ball of sticky seeds that tenaciously grip the fur of cattle and the socks of hikers, but disintegrate into individual barbed seeds when one attempts to pluck them off. *Acaena* is an introduced weed not native to New Zealand. This inconspicuous herb is an indication that the New Zealand Alps are not as pristine a wilderness as they seem. Introduction to New Zealand by European sportsmen of Middle Eastern fallow deer (*Dama dama*), Himalayan tahr (*Hemitragus jemlahicus*), and Australian brushtail possum (*Trichosurus vulpecula*) has been hard on native vegetation not conditioned by such animals. As native plants are trampled and overgrazed by large mammals, and tree canopies thinned by possums, New Zealand's diverse native bird fauna edges toward extinction. Bare patches of soil are ripe for colonization by introduced grasses and other weeds. These are changing the structure and texture of soils in the same way as a farmer improves pasture with new breeds of grasses. Even national parks and wilderness areas are modified by sometimes careless, and sometimes overprotective human management.

Current unwitting modification of the last bastions of wilderness is a new frontier in an ongoing process. More than 900 years ago, New Zealand Maori beached their canoes from Polynesia and introduced rats and dogs to these islands, which until then had no native mammals other than bats. They set fire to the forests and hunted the large flightless birds of New Zealand, the moas. In place of beech (*Nothofagus*) and podo (*Podocarpus*) forests, Maori promoted the spread of tussock grassland. All 11 species of large flightless birds (moa) were driven to extinction within only a few hundred years. The clayey subsoil and raw humus of primeval forest soils were lost to erosion and replaced with the fine, clayey humus of grassland soils. The golden native tussock grasses rippling in the wind are a beautiful sight, but a great fortune of soil capital was lost to Maori fire management. Maori populations stabilized at about the time of the Medieval warm period, when the great cathedrals were built in Europe. Maori abandonment of much of the deforested South Island to the slow and relentless humus accumulation of tussock grasses corresponds to the Medieval Little Ice Age, when oxen could be roasted in London on the frozen Thames River and wine-grape growing was abandoned in Britain. New Zealand's past history was linked to that of the rest of the world by global atmospheric change.

Mountainous New Zealand, with gravity pulling on all sides at its limited agricultural land, has a dwindling soil resource that strongly restricts its contribution to the global carbon budget. In contrast were the use, then abuse, of the soils of the vast plains of North Africa and eastern Austria, which fueled the rise, then allowed the fall of Ancient Greek and Roman Empires. These are the globally important exchequers of atmospheric carbon, the currency of the Proserpina Principle of biological cycles of climate.

Civilizations and soils gained and lost are most apparent in Mediterranean lands. Once a land of milk and honey, but now the abode of goats and dust, Israel was home to two of the world's three great religions. The mighty cedars of Lebanon were long ago cut for ships and towns of Phoenicia, where the alphabet was invented. The fragrant litter and clayey soils of the cedars were washed from rocky hillsides now terraced by farmers to conserve what little clay and humus remains. The wheat fields of North Africa were once the granaries of ancient Rome, the source of so much of our language and so many of our legal and commercial institutions. The deep, brown, grassland soils of North Africa were replaced by dunes and dissected by agricultural terraces. It has long been clear that soils, like civilizations, do not last forever. The carbon-rich soils of the Mediterranean Basin, like the glories of Greece and Rome, have long ago gone up in smoke.

In his *Critias*, Plato (ca. 429–347 BC) wrote of the fertile grassy landscape of his ancestors, compared with the rocky barren hills of Athens in his own time. He wrote (translated from ancient Greek) "in comparison of what then was, there are remaining only the bones of the washed body … all the richer and softer parts of the soil having fallen away, and the mere skeleton of the land being left." Greece was not always bright with harsh sunlight reflected from whitewashed houses and gray rocks. Some 10,000 years ago, it had herds of antelope and prides of lions in grassy woodlands of pine (*Pinus*), lentisk

(*Pistacia*), and evergreen oak (*Quercus*). The mountains of stone were then more like the kopjes of the Serengeti, mantled by red, clayey soils.

The loss of a bountiful biota and soil resource though time is recorded in many parts of Greece as sequences of red paleosols covered successively by layers of gravel, sand, and silt. A colorful record of red and yellow bands is exposed in numerous road-cuts, quarries and creek banks near the base of the hills. These records of five millenia of landscape change are well known in the Argolid, a rugged northeastern part of Peloponnesian Greece. Tjeerd Van Andel and Kevin Pope spent several summers there deciphering these records, returning to Stanford University for laboratory studies. The red bands are buried soils representing periods of landscape stability and life. The gray-yellow layers, on the other hand, represent times of destruction, when soil slumped or was blown from nearby hillsides and accumulated on the foot-slopes and valley bottoms as layers of sand and silt. These were times when an intricate mixture of minerals, microbes, and moisture, interlaced with roots and worms, was abruptly stripped by erosional gullies or smothered with mudflows.

The causes of these periods of soil erosion and mudflows interrupting longer periods of soil formation can be unravelled through careful study of the red-banded alternations of paleosols and sediments in gullies dissecting the footslopes of the hills. Potsherds and other artifacts buried within the sequences allow dating of these events in relation to growth and decline of human populations. Pollen grains preserved in nearby bogs reveal vegetation similar to different regions of Greece today: oak (*Quercus*), pine (*Pinus*), olive (*Olea*), lentisk (*Pistachia*) and grasses (Gramineae).

The greatest and most destructive of the pulses of soil erosion was during clearance of the land for agriculture some 4000–5000 years ago, when Early Helladic pottery was in vogue. Agriculture had already been practiced for five millenia before this in China and the Middle East, but expanded into Europe, Africa and Central America 5000 years ago at a time of marginally warmer climates. In Greece, the valleys were first cleared of trees, and planted with barley, wheat, and olives. Soon trees of the hillsides were cut for livestock forage and fuel wood. Stripped of trees and with grassy cover thinned by livestock, small channels cut by rain into the red earth grew to large erosional gullies. After heavy rains, large blocks of soil were transported down these gullies to form thick deposits of angular red clayey blocks in a muddy matrix at the foot of the slope. By 4000 years ago, the Argolid's population declined dramatically, to judge from the low density of artifacts and settlements in soil and sediment overlying the first great wave of mudflows. Cooling climates are then indicated by a rise of oak and pine pollen. In other parts of the world the pattern of population growth and decline was similar at a time when ice cores indicate that the whole world edged into cooler times. Careless squandering of the soil led to famine, emigration, and cooling.

The mudflows of 4000 years ago are capped by another red soil, also an Alfisol, but not so thick nor clayey as the primeval Alfisol. It takes time and the activity of millions of organisms to convert a silty or sandy accumulation of minerals to a working, breathing Alfisol soil of reddish clay and humus. Unlike the early woodland and herds of game, this was a new landscape of grazing livestock, olive groves, fields of grain, and agricultural weeds, which left pollen in coastal bogs of the time. The thinner, less clayey, brownish-red soil, reflects this human modification. No longer was the deep, red clay sucked dry of moisture by large roots of trees, but pasture grasses and grain crops formed new, thin, finely structured, brown surfaces on deeper red clayey portions of the old Alfisol soils. These new soils and seafood from Greece's ragged coastline were the economic base of the Bronze Age Mycenaean civilizations of Homeric legend. They fueled growth of the walled city-states of Mycenae and Tiryns, with their gigantic stonework (Fig. 3.2). Their walls and weapons tell of wars in a world of populations of unlimited ambition, but limited resources.

From this time forward, literature supplements the record from paleosols and artefacts with a wider view of human motivation. Homer's epic poem of their times, the Iliad, emerged from oral tradition some years afterward and taps rich veins of patriotic, mythic, and religious sentiment. Perhaps the Greek rescue of Helen from the Trojans was in defense of male honor, as Homer implies. Perhaps it reasserted a legal claim to the throne of Sparta in a society with matrilineal inheritance. But this and other sackings of Troy, documented by archeological excavations, have more prosaic explanations. Troy commanded an important trade route for horses, salt, spices, and gold through the narrow channel connecting the Black and Aegean Seas. The great agricultural societies of the age were in Egypt and the Middle East, which exported not only carbon to the atmosphere, but imperial ambitions to the petulant and arrogant heroes of the Iliad. They knew well the territorial ambitions of the peoples to the east, because even before Homer's heroes of Troy, Mycenae and Tiryns, these great cities had been conquered by Indo-European speakers of linear B, a version of archaic Greek. Homeric palace culture thrived during a warm climatic interval as an outpost of the great Egyptian and Babylonian engines of global change.

They did not last, because Kassites from the north sacked Babylon, and Akhenaten's religious revolution plunged Egypt into turmoil. Within a few decades of the conquest of Priam's Troy, the horsemen and pirates of Asia were

Fig. 3.2 The Lion Gate and Upper Grave circle of unroofed beehive tombs, at the Bronze age (3600 years ago) citadel of Mycenae, Greece. The beehive shape and long entry passage may have been a symbolic addition of a uterus to Mother Earth as part of an archaic goddess cult

on the move again, plunging Greece into long Dark Ages about 3400 years ago. Soils were principal beneficiaries of these Dark Ages when little was written or made. Human populations in the Argolid declined dramatically, to judge from scarcity of settlements and artifacts. For some 300 years of ancient Greek history, we have little historic record. By the end of the Dark Ages, Bronze Age goddess cults of "Our Lady" (Hera in Greek) finally yielded sway to the familiar classical pantheon ruled by Zeus. The ancient goddess and other local dieties were not forgotten but recast as his subjects: Athena, Hera, Aphrodite and the Harpes and Fates. This political and religious unravelling of the thread of Greek culture was not accompanied by profound soil erosion. In the Argolid, oak and pine regenerated from fallow fields and underexploited grazing lands with climatic cooling. Regenerating forests arrested soil erosion, deepened soils and enriched them with uninterrupted accumulation of soil humus from plant decay and weathering of mineral grains to nutrient-rich clay. The natural ability of plants and soils to recover once human demands are relaxed is both surprising and encouraging. It was a reversal of human soil abuse, that led to a reversal of centuries of climatic warming.

The growth of classical Athenian and Spartan city states 2500 years ago brought new pressures on the landscape and renewed warming, as documented in detail by Plato, Thucydides, and other classical authors. Forests were cut to rebuild towns sacked in war and fleets of war ships. Where there were once verdant cattle pastures, overgrazing left unpalatable and thorny weeds suitable only for goats. Rocky ridges and knolls emerged beneath eroding soil. The new soils formed on the deposits of mudflows and gully erosion were thin, brown Inceptisols, with less clay and iron oxides than the primeval Alfisol soils. Thin, rocky soils and weedy vegetation disturbed by expanded cultivation and grazing made for landscapes more readily perturbed by floods and storms. Floods and mudflows now followed storms of a magnitude that would have had little effect in earlier ages, and the products of these storms remain as the record of paleosols and mudflows in the gullied footslopes of the hills. Such changes are much more obvious around overpopulated Athens, compared with rustic landscapes of the Argolid, which is still a patchwork of more and less degraded soils. Intensive use of soils paid for the glory of Classical Greek art and philosophy, but as soil and forest resources waned, so did the political power of Greece. By some 2000 years ago, it was falling within the shadow of the vigorous new nation state of Rome. One expanding empire was supplanted by another, which by raiding the soil resources of North Africa and eastern Austria induced continued warmth and productive rains. Pax Romana of the Caesars stabilized the falling population of Greece.

By 1600 years ago, Greece was a rural backwater between rival seats of a Christianized but divided Roman Empire based in Milan to the west and Byzantium to the east. During the seventh and eighth centuries AD, Imperial edicts ordered destruction of icons of Jesus, Mary and the saints, which were venerated in the style of classical gods and goddesses. Local riots and defense of icons, among other religious controversies, eventually separated the Eastern Orthodox Church from Roman Catholicism. Religious factions soon led to less effective military and civil administration, declining urban services and plagues. Slavic raids depopulated Greece and other marginal regions of the late Roman Empire, where centuries of cropping resulted in declining crop yields. Once again, human population decline at a time of political and religious crisis enabled recovery of soil, vegetation, and building of soil humus. The fallow fields and hills, protected from erosion by the growth of weeds and scrub, again built reserves of clay, humus, and available-nutrients in reddish brown, clayey Inceptisol soils. Final decay of the Roman Empire coincided with climatic cooling that mobilized the hardy tribes of the north to move south to sack Milan, Byzantium, Athens, and even Rome itself.

The next wave of mudflows and gully erosion accompanied resettlement and deforestation of overgrown fields and grazing lands by twelfth century feudal barons from southern France. This Frankish expansion throughout the Mediterranean was at a time of warming climate when the great cathedrals of northern Europe were built. Bubonic plague, war and fires of the fourteenth century severely depopulated Europe, but gave soils a respite followed by a bitterly cold century of the Little Ice Age. Subsequent rebound of European population in the seventeenth century and global colonization in the New World, Africa and Australia broadcast landscape degradation and heralded global warming that is accelerating today. Landscape and soil changes were less marked in Greece, which was then at the neglected fringes of the Ottoman Empire.

The Argolid still has pockets of useful and even ancient soil, unlike Athens' barren, rocky ridges. Greece has a dry climate, but desertification of its soils has proceeded without obvious local change to desert climate or soils. Human modifications of its soils were not so profound as to push Greece into a different climatic zone. Its Inceptisol and Entisol soils are like those of deserts in their thin humus, silty texture, and common little-weathered mineral grains, but few of them have abundant soluble salts such as gypsum and very shallow horizons of calcium carbonate nodules like true Aridisols of deserts. In dry climates, soils tend to wash or blow away. Ancient Greek deforestation and agricultural expansion were especially destructive. Nevertheless, there were long periods of sustained agricultural production during the Bronze Age and Classical-Roman times, and soils recovered when human populations declined during political and religious crises. Soil conservation practices such as terracing and carrying soil back uphill, remain effective outside the sprawling suburbs of Athens. Small houses along narrow alleys in rural Greek villages surrounded by terraced fields demonstrate a continued respect for the soil and its bounty. General lack of field fencing and continued employment of shepherds has also done much to prevent soil erosion. Humans affect soils in ways that do not mimic natural ecosystems, but which nevertheless have great consequence for local and global climates.

Dry regions such as Greece were not alone in suffering under human occupation, and playing a role in global climate change. Humid regions of the world are vulnerable in different ways. Unlike soils of aridlands, many soils of humid climates are clayey, sticky and forested, so retain moisture and humus, and resist erosion. But other problems aggravated by human interference in humid regions can be equally devastating in the long term. Vulnerable nutrient elements include calcium, magnesium, potassium, and sodium, which are taken up by plants and leached from the soil by rain and groundwater. The more rain and the lusher the vegetation, the more rapidly soil becomes acidified by alkaline nutrient loss. This natural process converts nutrient-rich glacial dust and till to infertile quartz-rich soil over periods of tens of thousands of years. Human activities such as cropping hasten this process by promoting erosion of ploughed soil and dispatching nutrients with harvested produce to market. These nutrients starve the growth of plants which is at the very beginning of carbon accumulation in soils, followed by its partial rotting and incorporation as humus. Human diversion and elutri-ation of soil nutrients can limit the sponge-like capacity of soils to absorb atmospheric carbon dioxide.

The humid English Lake District, like rustic Greek landscapes, was an idyllic retreat of nineteenth century romantic British poets, William Wordsworth and Samuel Coleridge. The windswept crags, broad sweeps of purple heath, and green mossy bogs were envisaged as the epitome of untamed nature. But five millenia of environmental change preserved in peaty and silica-rich soils of the Lake District show that this National Park has not escaped human modification, or the wider influences of global climate change (Fig. 3.3).

Winifred Pennington Tutin of Leicester University has reconstructed the Lake District's ancient history by sifting through tiny fossil pollen grains preserved in Histosols, or what the locals call "peat bogs" (Fig. 3.4). Pollen grains dating back 8000 years reveal a landscape of oak forests in fertile clayey soils. At that time the highlands were not cultivated, but summer visitors quarried fine-grained tuffs of the rocky peaks as cores for hand axes later polished in the lowlands. Neolithic farmers cultivated a few sandy ridges, spreading pollen of agricultural weeds into uncultivated swampy bottom-lands. By 5000 years ago farming had expanded over most of the flatter ridges, and a fungal pathogen selectively removed elm (*Ulmus*) trees. Also to blame for elm decline were cropping of elm leaves as forage, and greatly increased soil disturbance by expanding populations in a globally warm climate. The carrying capacity of the landscape was limited, because only soils with a texture amenable to hand-hoeing for cereals were cultivated. These farm soils

Fig. 3.3 Langdale Pikes (mountains) overlooking orderly fields rimmed by stone fences and a Scots pine (*Pinus sylvestris*) in the Lake District of England. Hard volcanic tuffs of the Langdale peaks were quarried for Neolithic hand axes as long ago as 8000 years. The lowland clayey soils are now pastures enclosed by stone fences, and were cleared of forests only a thousand years ago, long after cultivation of silty soils on rolling ridges

were silty and sandy, and more easily eroded than clayey or peaty soils. Thus began a cycle of soil degradation and liberation of cationic soil nutrients: Ca^{2+}, Mg^{2+}, Na^+, K^+. Once freed of forest cover, the small amounts of clay and humus that stabilized these silty soils were oxidized into the atmosphere or washed and blown downhill, and with them went the mineral nutrition of the soil. As these nutrients were depleted, soil acidity increased to levels that could no longer support good yields of cereals. Exhausted upland fields were abandoned to grasses and sheep grazing. After overgrazing and fire, heather and bracken invaded increasingly acidic upland soils, leaving their pollen and spores in peat moss as records of these vegetation changes, which coincided with the post-Mycenean Dark ages of Greece some 3000–3400 years ago.

There was another expansion of agriculture during warm times of the Roman Empire, when a Roman garrison at Hardnott guarded trade routes from mountain farms to the coast. With decline of the Roman Empire, cooling and expansion of glaciers in northern Europe drove Vikings southward. By the ninth and tenth centuries, Viking raids along the nearby coast

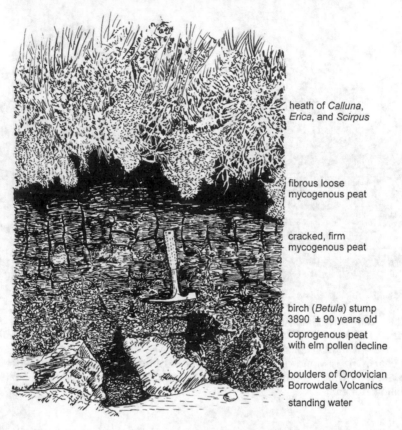

heath of *Calluna*, *Erica*, and *Scirpus*

fibrous loose mycogenous peat

cracked, firm mycogenous peat

birch (*Betula*) stump 3890 ± 90 years old

coprogenous peat with elm pollen decline

boulders of Ordovician Borrowdale Volcanics

standing water

Fig. 3.4 An excavation near Red Tarn, in highlands of the central Lake district, England. This sequence of peats and their pollen and fossil stumps in a Histosol soil, records the regional decline of elm (*Ulmus*) pollen some 5000 years ago, followed by local destruction of a birch (*Betula*) woodland, to create heath some 4000 years ago

initiated new migrations into hills of the Lake District, and many Scandinavians settled to farm the coastal plain. With rising temperatures and global trade in the seventeenth century, a new kind of plow was exported from China, first to Holland, and then to England. Its curved iron moldboard turned over the soil, rather than just furrowing it. This new plow enabled cultivation in clayey soils of valley bottoms of the Lake District, which pollen records and local folklore indicate were previously the domain of oak forests and wolves. Rising population, upland grazing, and lowland cereal production expanded throughout the region. Forest clearance led to soil erosion, indicated by clayey bands found within lowland peats. The soils of periodically waterlogged bottomlands thus suffered a similar fate to those of the previously cultivated uplands. Loss of nutritious clay and enrichment of infertile quartz in upland soils encouraged the formation of Spodosols, which are sandy soils with subsurface layers rich in rusty iron oxides. Spodosol soils are both highly acidic and low in fertility, because they have few of the alkaline elements found in clay and needed for plant nutrition. In waterlogged depressions of both uplands and lowlands, acidic and infertile soils support little more than *Sphagnum* moss, which accumulates and rots into a peaty blanket mantling the landscape, a kind of soil called a Histosol. Peat is the dark brown spongy precursor of coal, and can be cut and dried as a domestic fuel. It does not burn as hot or bright as coal, but can make cozy a moorland cottage. Peat cutting for fuel added more problems by exposing the underlying mineral soil to wind and water. The carbon stored in peaty soils was thus liberated to the atmosphere as nutrient-starved sandy soils of slow-growing vegetation were becoming more widespread, and climate became wetter and warmer.

With the advent of cheap imported cereals from British colonies in the nineteenth century, wheat growing was abandoned in much of the Lake District. By then, cereal yields were marginal, and sheep grazing continued to wear down the pastures. The mid-nineteenth century Lake District of William Wordsworth was already an acidified landscape of limited agricultural worth. There remains to this day splendor in grass and grandeur in the high rocky crags, but patchwork human disturbance has accelerated the natural progression of environmental acidification. The process began well before heavy industry entered the English midlands, and with it the spread of sulfuric acid from coal burning. Soil acidification by industrial pollution is merely the last phase of a centuries-old process. Soil carbon has been spent by human exploitation, and lofted into the atmosphere, just as effectively as during the desertification of Greece.

Although these case studies show many local and historical peculiarities, there are broad similarities and even some coordination of events in time. Expansion of empires and populations coincide with warming climates, and appear to induce warming by generating atmospheric carbon dioxide through burning of wood, peat and other fuels, and land clearance and plowing by expanding agriculture. At the same time, tilling and harvesting reduce soil thickness, texture and fertility so that it can no longer absorb atmospheric carbon dioxide through growth of plants and creation of humus. Conversely periods of population decline and strife were times when fallow soils accumulated carbon by reabsorbing atmospheric carbon dioxide with consequent climatic cooling. The rise and fall of empires thus coincides with the rise and fall of soil abuse and consequent increase in atmospheric greenhouse gases and temperatures. Temperature peaks do not coincide with critical imperial inventions, but rather with peaks of imperial expansion. Key inventions, such as Roman military centurions, or the moldboard plow, are causes rather than consequences of global climate change. Inventions were critical, but it was their widespread application that made a difference and enabled lasting monuments of pyramids, temples and cathedrals. The inventions and beginnings of empires did not come at times of climatic warmth, which were rather the culminations of empire and of human abuse of that greatest of all climatic regulators, soil. Many of my colleagues who spent months at sea in oceanographic investigations have urged me to consider the oceans as the principle climatic regulator. But oceans are big and inert, and like the flywheel of a car engine, are a force for homeostasis, redistributing heat around the world by means of ocean currents. They may blunt swings in climate generated by the mosaic of deserts and forests, but the source of climatic oscillations through time is changing soil.

Another intriguing result of these studies of soils and landscapes in collision with human needs is that human modification of the landscape is not entirely the fault of our generation. Undoubtedly, the pace of land degradation has accelerated as world population has exploded and as earthmoving technologies have improved. But soil has been modified for centuries, using fire at first to improve pasture for game, then increasingly destructive deforestation and cultivation. Much of the world's fertile land has long been under human control. As Bill Ruddiman would argue, agricultural expansion for the past 10,000 years has maintained climatic temperatures in a rough equilibrium despite changing solar and orbital parameters. We have inherited a system of environmental change that has such momentum that it cannot be reversed. Knowing the nature and the rate of change, we can perhaps mitigate the worst excesses of deterioration. Fertility can be restored to acidic soil by adding lime, bone, or other fertilizers. Desertified soils can be stabilized with nets, fertilized with humus, and watered. Cell grazing, contour coppicing and pasture cropping can restore soil moisture and humus to degraded pasture lands. What has been lost is a vast soil resource many thousands of years in the making. Soil resources ultimately control our tenure on this planet. The blowing away and washing downstream of the soils of the central North America and southern Russia are arguably our greatest environmental problem, more important than local acidification and desertification of long-abused lands, or geographic dislocations due to global warming and sea level rise. The North American Great Plains and the Russian Plain are the great bread baskets of the world. When these soils are depleted, humanity will have to change, as it did during the last great soil crisis and global change during deglaciation some 12,000 years ago. We are not the worst thing to have happened to soils in the geological past, as we shall see in following chapters.

THE GLORY OF GREECE

Greece of Plato was built not with words,

Rather with cattle and goats in herds,

With olive groves, and fields of grain,

Tree-clad slopes cut again and again,

Building towns sacked for plunder or spite

And ships made to show military might.

Kings prospered and kept household bards,

Telling more of the past than pottery shards.

But immortal lays were bought at great cost,

Of slopes without trees and thick soils lost

Downhill. From woods and thorny heath

Rose crags as welcome as wolves' teeth.

4

Humanity from Global Change

Before extensive agricultural modification of soils, human evolution, like that of other animals, was shaped by environmental changes. Paleosols at key sites for human and pre-human fossils reveal the role of climate and vegetation, particularly grasslands and forest, in human evolution.

Soils develop through time, to rhythms like those of daily life, as well as to longer rhythms that have seen the rise and fall of civilizations. The daily rhythm of morning dew and midday sun moistens and then dries the cracks and surface of the soil. Evaporating dew adds to the invigorating chill of morning, whereas bright sun glares back from dry soils and leaves at noon. Annual rhythms of spring rain and fall drought, summer heat and winter chill, also cue the activity of worms and roots, and the opening and closing of deep cracks, in which water and air weather the mineral grains of soil. The chill of late winter is lessened by rising levels of carbon dioxide from soil respiration, just as the heat of late summer is lessened by the draw-down of this greenhouse gas into the leaves of trees. This is the very time scale at which the Proserpina Principle of cooling plants and warming animals is best demonstrated by a half century of measuring the ups and downs of atmospheric carbon dioxide (Fig. 4.1A). These measurements also reveal a continued rise in carbon dioxide and global warmth over the past century as humans have commandeered most of the resources of our small planet.

Can the Proserpina Principle of oscillations in plants and animal populations also explain long term climatic oscillations of the past Ice Age, and the

© The Author(s), under exclusive license to Springer Nature Switzerland AG 2022
G. J. Retallack, *Soil Grown Tall*,
https://doi.org/10.1007/978-3-030-88739-1_4

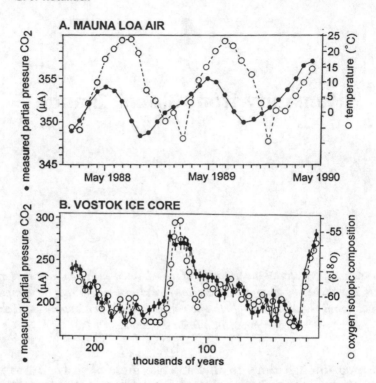

Fig. 4.1 Variations through time of atmospheric carbon dioxide content on annual and millennial time scales. Annual variation measured at Mauna Loa Observatory, Hawaii (**A**), is only three months out of phase with temperature, and is caused by spring draw-down of carbon dioxide by photosynthesis and fall rise of carbon dioxide due to soil respiration. Multimillennial variation in carbon dioxide abundance measured from air bubbles in long cores of ice from Vostok, Antarctica (**B**), is in phase with temperature changes that can be inferred from isotopic composition of oxygen in the same bubbles. The big spike in this graph corresponds in time to the warm-wet climate of the Sangamon paleosol, and smaller spikes with the Farmdale and Gardena paleosols

evolution of human species? These are time scales of soil initiation, development and destruction, which coincide well with cycles of climatic change. Evolution and extinction of our evolutionary ancestors also was cued to these climatic cycles, as we evolved from a minor player among African animals to a major global force for change. Our evolutionary tree has been severely pruned by past global changes, and we should be concerned lest our surviving branch is lopped off by future changes. Sequences of paleosols remain as evidence of environmental change during the evolutionary career of our ancestors. Furthermore, paleosol archives are so similar in such different parts of the world as Illinois, Germany, Argentina, Tanzania and China, that some of these climatic events were globally synchronous. We are now in the midst of global change induced by human consumption of soil and other resources. Big changes are in store for us, but this is not the first time creatures like us have faced global climatic challenges.

Picturesque evidence of past climate and ecosystem shifts comes from paleosols. A brown stripe running through creek banks and road cuts of northern Illinois, called the Farmdale paleosol (Fig. 4.2), represents a respite from the frigid gales that deposited windblown glacial dust and formed thin, ice-cracked, peaty paleosols (Gelisols) before and after formation of this paleosol. The Farmdale paleosol is clayey and brown, an Inceptisol like current forest soils of northern Canada. In places the Farmdale paleosol has yielded fossil cones of spruce (*Picea*) and pine (*Pinus*). Similar soils, now represented by the Poperinge, Hoboken, and Zelzate paleosols, were widespread some 40,000 years ago in northern Europe. These paleosols represent abrupt transitions from boreal frigid desert to taiga forest, followed by long-term cooling to glacial conditions about 25,000 years ago.

The most reliable records of this climatic warming 40,000 years ago come from chemical analysis of atmospheric gases such as carbon dioxide and oxygen in tiny bubbles buried deep within the ice caps of Greenland and of Antarctica (Figs. 3.1, 4.1B). These antipodal ice cores contain near-synchronous records of this warming, indicating its global extent. Warming from the coldest temperatures of the past 100,000 years was completed within only a few decades around 40,000 years ago. This time of warmth lasted only a few thousand years, before temperatures again dropped, gradually but with continuing fluctuation, over the ensuing millennia.

The extreme climatic volatility of the Ice Ages of the past 1.6 million years is commonly attributed to variations in inclination of the Earth's axis of rotation and in eccentricity of its orbit, which affect the amount of solar radiation intercepted by the Earth's surface at high latitudes. Such astronomical forcings are named after Milutin Milankovitch (1879–1958), a Serbian scientist who first made the link between these astronomical variations and paleoclimate. It is surprising that such minor astronomical forcings could result in climatic swings big enough to bring down large ice caps over Europe and North America. Furthermore, the near-synchrony of paleoclimatic change in the northern and southern hemisphere is a puzzle for models involving the orbital variations responsible for precession of the equinoxes. This variation in orientation of the Earth's rotational axis determines where summer and winter begin within the elliptical orbit of the Earth around the Sun, and so should favor one hemisphere over the other. However, climatic change in both hemispheres is near-synchronous. Also puzzling is new dating of the last four big climatic shifts which did not occur exactly every 100,000 years as predicted by Milankovitch astronomical forcing, but varied in length from 80,000 to 125,000 years. Finally, a really odd feature of the climate shifts is that ice core and other records show a pattern of long-term cooling punctuated by sudden warmings called terminations. Milankovitch astronomical forcings have the symmetrical undulating pattern of ripples on a pond, but actual records show a jagged, asymmetric, saw-tooth pattern of temperature through time, like a system driven to breaking point. These discrepancies indicate that astronomical forcings are not the whole story. Ecosystems and their soils also had a role in amplifying climatic changes, because each climatic swing involved profound changes in vegetation that in turn affected the balance of photosynthesis and decay, of growth and burial, of plants and animals.

Buried soils such as the Farmdale paleosol represent long-term development of ecosystems, then an abrupt breaking point. The brown clayey Farmdale paleosol is separated from the black grassland soil (Mollisol) of the past 12,000 years by a layer of light gray silt, of the distinctive kind called loess. The fresh loess minerals delivered by wind are the beginnings of soils, their parent material. Plant nutrient elements such as calcium and potassium are unlocked from mineral grains by mildly acidic solutions of carbonic acid derived from dissolution of the greenhouse gas carbon dioxide within soil water. The loss of nutrient elements to roots and consequent building of plant biomass is slow, but relentless. It steadily consumes carbon dioxide and other greenhouse gases to produce life and soil. Over time, mineral nutrient uptake and productivity wanes as soil feldspars and other minerals are converted to less fertile clay and iron oxides. As soils become depleted of nutrients their vegetation thins and becomes more susceptible to fire, insect pests, and storms. Eventually the ecosystem burns, rots, or is torn out, and liberates its carbon to the air. The long period of carbon fixation and plant growth that takes carbon dioxide from the air, is followed by overgrazing and dust-bowl conditions of increased carbon dioxide release from soils. Such a cycle of atmospheric change is like the yearly Proserpina Principle, but on millenial time scales. One could argue that such long-term soil cycles are cued to Milankovitch astronomical forcings, like the winter and summer variation in sunlight received on Earth. But long-term variations in radiation are too slight to completely explain the large temperature changes observed, and call for amplification by terrestrial changes in soil, vegetation and atmospheric composition.

By this view, 12,000 years ago was a turning point, a catastrophic warming of sparsely vegetated loess soils that had accumulated from millennia of dust storms, advancing glaciers and frigid temperatures. Within the space of a human lifetime, the old life-ways of the tundra were swept away for a new opportunities of grasslands and game. Native Americans already in Oregon 15,000 years ago after coastal migration from Siberia, expanded their range into grasslands and aridlands rich in megafauna. The most striking warming of 13,200–12,900 years coincided with big game hunters using fluted Clovis points. The three-century success of Clovis technology may account for extinction of mammoth and other megafauna at this time, after which the big Clovis blades were abandoned for a kit of smaller stone tools. Extinctions were not due to the dramatic climate change of this time, because many equally dramatic glacial terminations earlier in the Ice Age were not accompanied by extinction.

Some 40,000 years ago was a minor warming enabling northward expansion of taiga forests and formation of the Farmdale paleosol on earlier loess of Illinois. No humans are known from North America at that time, but in southern Europe at this time, finely worked, specialized arrow points, traps, digging sticks, and gathering bowls were created to exploit berries and small game, by a new people called Cromagnons (*Homo sapiens*), invading Europe from the south and displacing formerly resident Neanderthalers (*Homo neanderthalensis*). With these newcomers came the earliest evidence of human artistry in caves of the limestone country on the Pyrenean slopes and plateaus of southern France and northern Spain. Giant ocher and charcoal images of bison, horse, and other animals were painted on cave walls here as early as 40,000 years ago. Rock art of similar age and comparable themes is known in the Sahara, East Africa, and Australia. Some Western Australian, circular, rock engravings have been thought to be 60,000 years old, although a recent reevaluation makes it unlikely that they are significantly older than the great cave paintings of Europe. Why was there such a burst of creativity all around the world at this time? There was a crisis of global warming and shifting habitats, but also fertile new ground for growth, procreation, and even leisure.

Even further back in time, some 130,000 years ago, the world lurched rapidly to another greenhouse peak from the depths of glaciation, as indicated by an inflection in chemical composition of polar ice cores as dramatic as the deglaciation of 12,000 years ago. Evidence of this warm time is as striking as a bold red line through homework. In the creek banks of northern Illinois, gray glacial silt and boulders overlie 4–5 ft. (1–2 m) of red clay, hackly with ancient soil cracks and piped with irregular traces of ancient roots. This older and more deeply buried soil has been called the Sangamon paleosol (Fig. 4.2), and is comparable in many ways with Alfisol soils under oak-hickory forests in Illinois today. Similar modern-appearing paleosols formed at about the same time in Europe, where they are called Rocourt or Eemian paleosols, and in the loess plateau of northeast China, where the paleosols have been

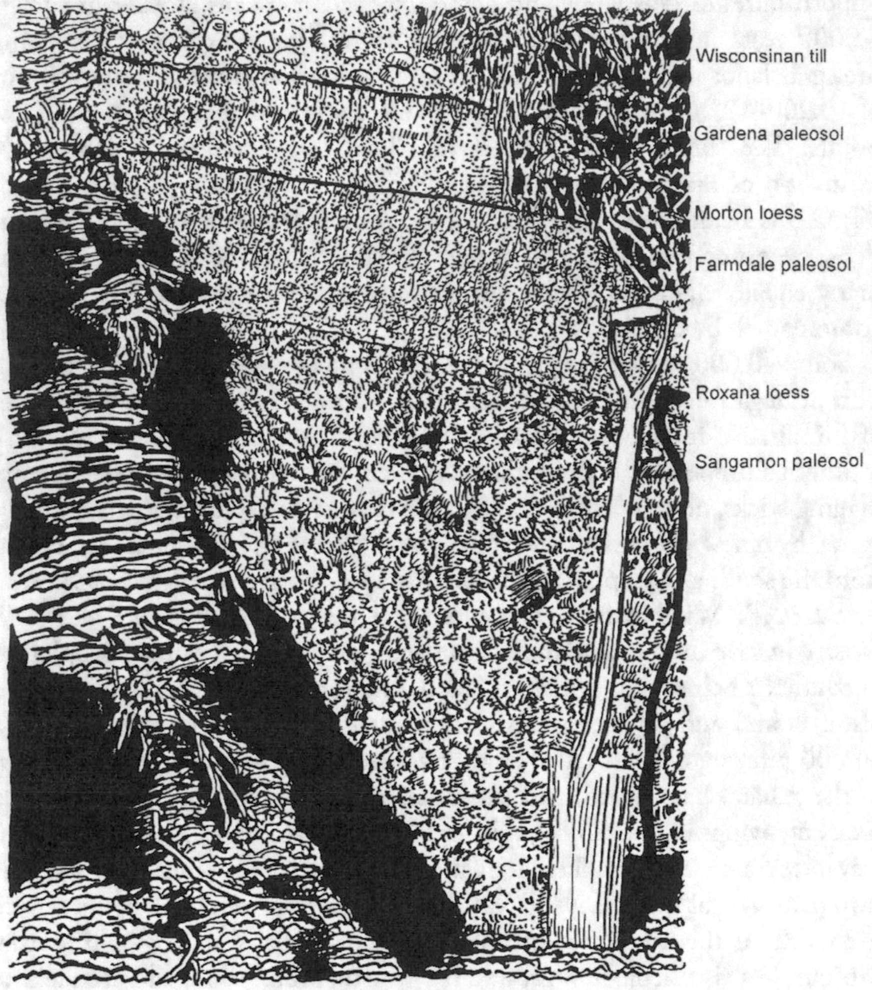
Wisconsinan till
Gardena paleosol
Morton loess
Farmdale paleosol
Roxana loess
Sangamon paleosol

Fig. 4.2 Excavation of the Sangamon paleosol (130,000 years old), Farmdale paleosol (40,000 years old) and Gardena paleosol (20,000 years old), beneath boulder clays deposited during the last glacial advance, in the steep banks of Farm Creek, near Peoria, Illinois

given the prosaic label of S1. Some 130,000 years ago, the magnitude of the chemical shift in the ice cores was much greater than at 40,000 years ago, and so too was the shift from ice and tundra to forest communities and soils comparable to those of today.

The glacial termination that ushered in a Sangamonian greenhouse spike may also have enabled modern humans, *Homo sapiens*, to prevail over earlier human species such as *Homo heidelbergensis* and *Homo antecessor*. We differ from more archaic varieties of humans in our strong chin and capacious, balloon-like cranium, distinct from the ridged brow and retreating jaw of earlier species. The origins of modern humans can be traced to Africa, where the earliest of these jutt-jawed and big-brained skeletons are found in deposits roughly 250,000 years old. Climatic terminations were times of profound dislocations and new resources, when some populations made great gains. Early humans evidently found the woodlands and forests of Europe and Asia to their liking, spreading out of Africa through a Middle Eastern corridor that was less arid and warmer than before.

There is another line of evidence for the origin of modern humans from studies of the molecules of heredity, and interestingly, it agrees broadly with evidence from fossils. The complex molecule deoxyribose nucleic acid (DNA) is found not only in the nucleus of cells, but in the small power plants, or mitochondria, of animal and plant cells. Mitochondria are scattered in the groundmass of our cells, and are not involved in the exchange of nuclear DNA between egg and sperm that creates a new human baby. Mitochondrial DNA is inherited only from our mothers. Studies of the variation of mito-chondrial DNA could have been done from a variety of tissue samples, but placentas proved to be a convenient large sample of tissue from living women who could be interviewed about their racial heritage. Comparison of the detailed sequence of component bases in mitochondrial DNA of thousands of women of different human races has shown that the greatest diversity and most primitive sequences are from Africa. Rates of change of this complex molecule also are known, though not as precisely as we would like, and calcu-lations give ages of 130,000–290,000 years for the African woman who was the earliest common ancestor of all the races of modern humans. There is remarkably little trace of earlier genetic sequences. It is as if an African race of modern humans spread through Eurasia, supplanting prior genetic types, beginning at least 130,000 years ago, when paleosols indicate global warmth and widespread woodlands at high latitudes. The totality of the genetic shift and the wide dispersal of modern human skeletons is evidence that this conquest was swift on geological time scales, and perhaps also brutal. At this time of global change, our species of humans found opportunity.

Even further back in time, some 1.2 million years ago, another landmark in human evolution stands out at a time of profound change in soil and climate. An abrupt termination to greenhouse conditions of 1.2 million years ago is marked by a prominent band of red paleosols within Olduvai Gorge, Tanzania (Fig. 4.3, Color Photo 4.1). These red paleosols are Inceptisols with common, large, calcite-encrusted, root traces. The red paleosols in Olduvai Gorge have been labelled Bed III. In Europe, this greenhouse termination is represented by the Turnhout paleosol and Waalian Interglacial. In Europe,

Naibor Soit

Ndutu Beds

red paleosols of Bed III

gray paleosols of Bed II

Fig. 4.3 The Castle, a mesa of red calcareous paleosols in central Olduvai Gorge, Tanzania. Underlying these dry woodland paleosols 0.8–1.2 million years old are gray and brown grassland paleosols yielding important human ancestor fossils as old as 1.8 million years

Color Photo 4.1 Pleistocene (0.8 to 1.2 million years) red Aridisols of Bed III, above gray Aridisols and Vertisols of Bed II in Olduvai Gorge, Tanzania

greenhouse warmth was a critical factor in forest expansion. But in Africa, the greater moisture of greenhouse climates was probably more important to the spread of woodlands. These red paleosols underline another pruning and simplification of the human evolutionary tree (Fig. 4.4).

Olduvai Gorge was made famous by Louis and Mary Leakey and the fossils of human ancestors they found there. Louis knew they would be found eventually from his first visit in 1931, when he discovered abundant stone tools littering the badlands. He was a man not known for patience, but it was not until 1959, some 28 years later, when Mary Leakey was on a walk with her dalmatians and Louis was in bed with flu, that she found the first skull of a kind never seen before from gray paleosols beneath the red band. The skull had flaring cheeks and a prominent ridge running over the top. Its thick jaws were massive and its molars as big as coat buttons. Mary and Louis nicknamed their skull "the nutcracker," and gave it the scientific name of "*Zinjanthropus*" *boisei* or "Boise's man from East Africa."

The first part of the name did not stick. It was soon realized that this fossil was closely related to a fossil previously found in a South African limestone quarry. The earlier discovery was found by a local farmer at Taung, but Raymond Dart, an Australian anatomy professor at the University of Witwatersrand in Johannesburg, South Africa, realized its significance as intermediate between ape and human. Dart named this fossil *Australopithecus africanus* or "southern ape from Africa" in a scientific article published in 1925. Dart's generic name *Australopithecus* preceded, and should be used instead of, "*Zinjanthropus*". There is also some support for using Robert Broom's older name "*Paranthropus*" for the robust lineage of australopithecines. The South African fossil is a juvenile skull, but so distinct from Mary Leakey's bruiser that it represents a separate evolutionary lineage.

Subsequent discoveries have shown several species of these two distinct lineages of early human ancestors. The robust lineage was big-boned and toothy (*Australopithecus aethiopicus, A. robustus,* and *A. boisei*). The gracile lineage was slender and petite (*Australopithecus africanus, Homo habilis,* and *H. rudolfensis*). Both lineages are represented by fossils of *A. boisei* and *H. habilis* in paleosols below the prominent red band of paleosols known as Bed III in Olduvai Gorge. When two closely related species of animals live in the same region, it is usually because they have different life styles, such as the pursuit predation of lions and arboreal ambush of leopards. The partial skeleton of *Homo habilis* discovered in Olduvai Gorge, is from an area of brown, crumb-textured paleosols, like Mollisol soils formed under wooded grasslands. Mary Leakey's original skull of *Australopithecus boesi*, on the other hand, comes from weakly developed paleosols studded with nodules

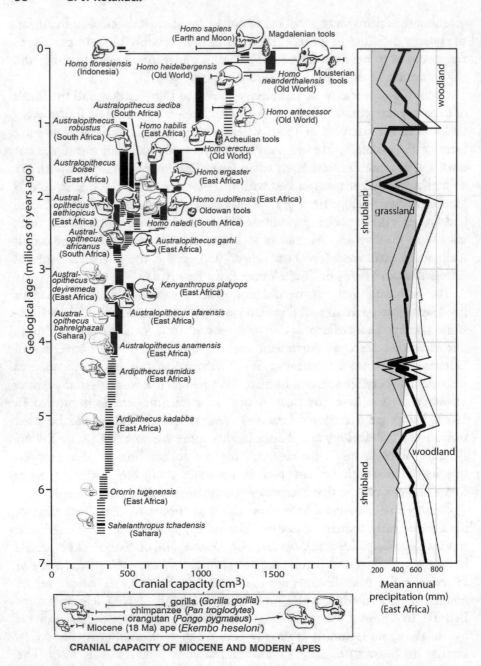

Fig. 4.4 A human evolutionary tree shows a great diversity of ancestors, here arranged in time and on an axis that reflects increased brain volume. The range of brain volume of some living and fossil apes is included in a separate panel below. The climate curve and vegetation is from paleosols in Kenya (reprinted from Retallack 2019, Soils of the past, with permission from J. Wiley and Sons)

of calcium carbonate, like Aridisol soils found under scrubby vegetation that can tolerate salt flats around seasonally dry East African lakes. More such observations are needed to confirm habitat preferences of gracile versus robust australopithecines, but their paleosols indicate they lived in a complex mosaic of riparian woodlands, wooded grassland, and lake-margin scrub.

Both *Australopithecus* and the earliest *Homo* skulls were small brained. Furthermore, they are associated with the crudest of tools, rounded pebbles with one chipped side of the Oldowan tradition, named for Olduvai Gorge. In addition, the fossil knees and hips of these early human ancestors indicate that they habitually walked erect. This was not what the doyens of British anthropology wanted to hear when they rebuffed Dart during a visit to London with his Taung skull. Misled by a clever fossil forgery including fragments of chimpanzee jaw and human skull bone supposedly from Piltdown in England, Sir Arthur Keith of the British Museum of Natural History thought human ancestors should have apelike teeth but big brains. The Piltdown fossils were widely acclaimed until they failed critical tests of dating and preservation in the 1960's. Only recently has it been learned that the forger was probably Martin Hinton, a part-time curator of the Natural History Museum. His motive may have been bitterness over denied vacation pay. Unluckily for Dart, the Piltdown fossils seemed to offer confirmation that it was brains that made us human. The Piltdown fossils tapped deep prejudices that intellect enabled our ancestors to rise up on their hind legs. Dart's idea of an erect-walking ape was an affront to this comforting idea. Like Pygmalion, scientists too become enamored with their creations. It takes evidence of unusual quality and quantity to unseat a favorite hypothesis.

The big shift in brain volume from some 800 to some 1200 cubic centimeters was called the "cerebral Rubicon" by anthropologists educated in British private schools on the Latin of Julius Caesar, who crossed that Italian river in defiance of the Roman Senate in 49 BC. It was *Homo erectus*, not Piltdown man, who crossed the brain-volume threshold, and brought a new technology of stone tools that litter the red paleosols, above the brown paleosols in Olduvai Gorge. Big game hunting was probably a motivation for manufacture of these new tools, which provided both large, leaf-shaped bludgeons and small, curved, cutting blades. *Homo erectus* also expanded beyond the confines of Africa, populating most of the Old World; Africa, Asia from Spain to China and south as far as Indonesia. Another similar species *Homo ergaster* may have migrated into Eurasia 600,000 years earlier, but records are scanty. The domineering population and culture after 1.2 million years ago was *Homo erectus*.

The red band in Olduvai Gorge represents global warmth and humidity of 1.2 million years ago (Fig. 4.3, Color Photo 4.1), and is a succession of buried soils with silty texture, calcite-encrusted root traces, and calcium-rich nodules shallow within the profile, as in Inceptisol soils of dry woodland. Below it are a variety of paleosols: (1) gray to brown, crumb-structured, clayey paleosols (Mollisols) of grasslands; (2) brown, blocky-structured paleosols (Inceptisols) of riparian woodland, and (3) yellow, platy-structured, salty paleosols (Entisols) of lake-margin salt scrub. This diversity of soils supported a diversity of early human ancestors, including *Homo habilis* and *Australopithecus boisei*. In contrast, the uniform woodlands inferred from paleosols in the red band supported only *Homo erectus*. Reduced heterogeneity of vegetation coincided in time with reduced early human diversity, and this time *Homo erectus* found opportunity.

Alternate sprouting and pruning is a marked feature of our evolutionary tree (Fig. 4.4). Early *Homo sapiens* emerged as the dominant human ancestor of warm climate forests some 130,000 years ago, replacing a variety of archaic humans of cold climate grassland mosaics. *Australopithecus africanus* emerged as the dominant species of woodlands some 3 million years ago, replacing at least two earlier erect-walking primates of grasslands. Our evolutionary tree appears to have diversified in the varied grassy habitats of cool glacial times, then been pruned back to one species in the monotonous woody vegetation of warm times. It has taken a while to discover this pattern, because human fossils are rarer than bones of carnivores, which themselves are less than 2 percent of mammal fossils. Our early ancestors were a minor part of African ecosystems and of the global carbon cycle. At this early stage in our evolutionary career, our ancestors were more likely victims rather than agents of global rhythms of soil and climate change.

Now let us step back to 4–8 million years ago to consider the evolutionary divergence of apes and humans at yet another global change for climate and soil, revealed by the paleosol archive of Lothagam Hill, within the desert of northwestern Kenya (Fig. 4.5, Color Photo 4.2). My own expedition of 1997 sought its shady canyons and oases of Nawata grass (*Cynodon dactylon*) and Doum palm (*Hyphaene thebaica*) after a long drive in the glaring sun of dunes and scrub in the untamed surrounding Turkana desert. It was a memorable struggle with flooded streams and muddy roads, which thwarted our earlier approaches. The creek crossing in front of a wall of onrushing turbid water was a narrow escape. The maze of rough bush roads, oppressive heat, and local bandits made us apprehensive. We felt privileged indeed to be admitted at last to the paleosols of these cool canyons and nearby badlands. The paleosols did not disappoint, as they change in texture and color from silty and

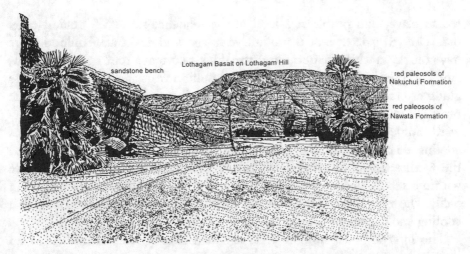

Fig. 4.5 Paleosols within a sequence of rocks 4–8 million years old, now tipped by mountain building so that the formerly horizontal rock layers dip to the right, in the canyons of Lothagam Hill, an oasis within the desert of northwestern Kenya

Color Photo 4.2 Late Miocene (6 million years) Kabisa paleosols (sandy Entisols), in the Nawata Formation, overlain by Apak member of Nakuchui Formation (Pliocene 5 million years) with Emunen paleosols (red Inceptisols), and Akimi paleosols (red Alfisols) below capping Lothagam Basalt (Pliocene 4 million years) at Lothagam, Kenya

red to clayey and purple at a level in the sequence some 5.5 million years old. The red paleosols of 6 million years ago and again at 5 million years have blocky cracking patterns and the thick calcareous root traces of woodland Inceptisol soils. The paleosols of 5.5 million years ago have smaller-scale granular cracking patterns and calcareous nodules at shallower levels in the profile, indicating drier, more open grassy woodland than before or afterward. Some 5.5 million years ago was a global climatic change, recorded as cooling that created glacial advances in Antarctica and drying that turned the Mediterranean Sea into a salt pan. The Mediterranean salt deposits are within a subdivision of geological time called the Messinian, after a town in Sicily. The term Messinian is now widely used for this time of dramatic global cooling and drying.

The interval 5–10 million years ago had already been identified as an important one for human evolution from a completely independent line of evidence. The study of DNA sequences of apes and humans, point to this as the time when the human evolutionary lineage diverged from that of the ancestors of modern African apes, chimpanzees and gorillas. Intelligence has been considered the hallmark of humanity, but Jane Goodall's work in Tanzania has shown that wild chimpanzees have impressive skills for problem solving and communication. The most fundamental anatomical and behavioral difference between us and apes is that we habitually walk erect and they do not. Walking erect is not a biologically simple affair, because it introduced new problems for childbirth and injuries of the back and knees. With such obvious drawbacks, there must have been compelling advantages for the evolution of erect stance.

The nature of these advantages has long been a cause for speculation, and there are three main competing ideas. One idea is that erect stance freed the hands for throwing stones or other weapons. This martial hypothesis was envisaged as an adaptation to life in the fiercely competitive world of lions and cheetahs in African savanna grasslands. Another idea is that human erect stance was an opportunistic adaptation to newly appearing grasslands, freeing hands for gathering small grain, standing tall to spy enemies over tall grass, or exposing a smaller surface area to the burning tropical sun. A final idea is that erect-walking humans used their hands to provide for helpless infants in a home base. Of the three hypotheses, this nurturing one does not treat erect stance as a consequence of grassland evolution. These family-oriented adaptations could have occurred in any environment, including woodland and forest.

We went to Lothagam Hill to evaluate these theories for ape-human divergence, by looking at the paleosols there for evidence of the kinds of environments in which our lineage evolved. There is a jaw fragment of a human ancestor, probably *Ardipithecus*, from a red woodland paleosol in that part of the Lothagam sequence dated to around 5 million years ago. Looking further afield, several skeletal fragments of *Ardipithecus ramidus* some 4.4 million years old from near Aramis, Ethiopia, and some 5.5 million years old from Alayla, Ethiopia, have thin tooth enamel and other skeletal features of chimpanzees. *Ardipithecus* probably walked erect, because its backbone attaches to the base rather than the rear of the skull. The fossils from Aramis were found in red paleosols with large root traces and nodules of calcium carbonate, like soils of dry woodlands. Associated fossil remains of woodland creatures, including *Colobus* monkeys, support this conclusion. Of comparable age at 4.2 million years old, *Australopithecus anamensis* was found by Meave Leakey's team in paleosols near Kanapoi, northwestern Kenya. These remains include outwardly flaring femur bones as evidence for the wide hips of erect stance. Most of the human ancestor fossils from Kanapoi were found in paleosols with the dark color and crumb-structure of grassland soils (Mollisols), but others were found in weakly developed paleosols with stout root traces (Entisols) that probably represent streamside woodland. From 5.5 to 4 million years ago, there were at least two species of erect-walking human ancestors, *Ardipithecus ramidus* in lake-margin woodland and *Australopithecus anamensis* in wooded grassland. The pattern of several human ancestors coexisting in grassland mosaics is at least as old as Messinian.

The diversity of erect-walking human ancestor fossils found in grassland-woodland mosaic after 5.5 million years can be contrasted with those from rocks some 6 million years old within northern Kenya's Tugen Hills. Martin Pickford and Brigitte Senut, both based in Paris, France, have named these fossils *Orrorin tugenensis*, which means "original man from Tugen" in a combination of Latin and local Tugen languages. The fragmentary fossils include teeth, jaw fragments and limb bones. Prominent articulating busses on fragments of femur are evidence of erect stance. *Orrorin* was found with a suite of red, calcareous paleosols and with fossils of forest mammals including *Colobus* monkeys. Another important fossil from the desert of Chad dated to about 6 million years old is Michel Brunet's magnificent skull nicknamed Toumai ("hope of life" in the local Goran language), or more formally, *Sahelanthropus tchadensis*. This skull was found with remains of fish, crocodiles, hippos and colobus monkeys suggestive of a lakeside gallery woodland. In other parts of the sequence, fossil soils contain nests of dung beetles and termites like those of open grassy woodland, which also is compatible with a different assemblage of antelope fossils. On present evidence though, it appears that erect stance in *Sahelanthropus* and *Orrorin* evolved at least 6 million years ago, and in woodlands rather than grassland mosaics. This line of evidence suggests that the evolutionary divergence of chimpanzees and the human family tree was not so much a matter of new grassland environments or technologies, but the need for hands to care for large and helpless infants. Paleosols add important pieces to this long-standing scientific puzzle as evidence for the vegetation where the bones accumulated.

Now let us step back 15 million years to another global change in soils and climate at yet another milestone in our long evolutionary history, the evolutionary divergence between apes and monkeys. This great evolutionary divide separates gorillas and chimpanzees on the one hand, from baboons, vervets and guenons on the other hand. Global climate change at that time is called the Monterey event, after the petroliferous Monterey Shale of California. The climate shift first inferred from study of foraminiferal microfossils in the Monterey Shale was a cooling and drying at about 15 million years ago from an unusually warm and wet greenhouse of about 16 million years ago in the middle of the Miocene epoch of geological time. This global climatic warm spike is recognized from the brief appearance of lateritic paleosols (Oxisols) in southeastern Australia, Oregon, and Germany, *Liquidambar* trees in Alaska, and large, tropical foraminifera in South Australia, as well as from changing paleosols in southwestern Kenya. Between 16 and 15 million years ago in Kenya red forested paleosols gave way to brown grassland paleosols, climate became cooler and drier, and the first monkeys of dry open grassland mosaics evolved from local forest apes.

The best known Kenyan sites for fossil apes and monkeys are around Lake Victoria, in the southwestern corner of that country. Fort Ternan is a large fossil quarry in hilly country west of the lake, but Maboko and Rusinga Islands, with their important fossil sites, are in the lake itself. For many years, those indefatigable fossil-hunters Louis and Mary Leakey took family vacations to these fossil sites. Young Richard Leakey, their son, was so bored by these working holidays that he refused to follow his famous parents into paleontology, pursuing a career as a tour guide and bush pilot. Ironically his aerial survey of badlands beside Lake Turkana led to expeditions and discovery of a fossil skull that made him also a household name in paleontology. Now called *Homo rudolfensis*, the skull was assembled from many small pieces by Richard's wife, Meave Leakey (Fig. 4.6, right). At about 1.8 million years old, it remains one of the most ancient skulls with the flat face and large brain that has come to characterize our branch of the human evolutionary tree. Richard's mother, Mary Leakey, discovered a comparably spectacular skull on Rusinga

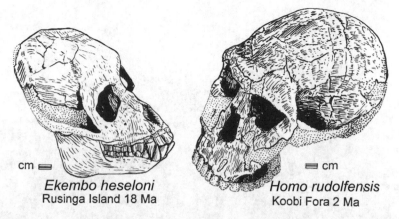

cm ▭
Ekembo heseloni
Rusinga Island 18 Ma

▭ cm
Homo rudolfensis
Koobi Fora 2 Ma

Fig. 4.6 Two famous skulls found by Leakey family expeditions in Kenya, contrasting the small brain, long snout, and prominent canines of an early ape (left), with the large brain, flat face, and reduced canines of an early human ancestor (right). *Ekembo heseloni* (left) was found on Rusinga Island by Mary Leakey during her husband Louis' expedition of 1948. *Homo rudolfensis* (right) was found in many pieces at Koobi Fora by Bernard Ngeneo on Richard Leakey's expedition of 1972, and restored like a three dimensional jigsaw puzzle by Richard's wife Meave Leakey

Island in 1948 (Fig. 4.6, left). Now named *Ekembo heseloni*, Mary's skull is 18 million years old, and is clearly an ape rather than human because of its elongate snout, small brain, apelike molars and prominent canine fangs. The body of *Ekembo* reconstructed from skeletal remains is quite unlike that of a chimpanzee or gorilla. Even though it had apelike teeth and lacked a tail, the rest of its body was more like that of a vervet monkey (*Cercopithecus aethiops*). *Ekembo* probably scampered about in the trees after fruit and leaves, ecologically more like a monkey than an ape. There were at least a half dozen species of these early monkey-like apes on Rusinga Island, some with longer arms than *Ekembo*, some larger and some smaller. The paleosols in which they are found on Rusinga Island include a variety of red, silty profiles with calcified roots and nodules like Alfisol and Inceptisol soils of a variety of woodlands growing along rivers, ephemeral streams and floodplains. The volcanic ashes of Rusinga Island also have preserved fossil leaves, insects and spiders, which confirm woodland habitats.

A whole new suite of soils appeared in southwest Kenya with global climatic cooling some 15 million years ago, and is best known on Maboko Island, also in Lake Victoria. These paleosols are brown, finely cracked into crumb clods and riddled with slender root traces like Mollisol soils of grasslands, but also have common lumps and rinds of black iron and manganese oxides as found in soils that are seasonally flooded. Dambo is the African name for those grasslands that turn to lakes and marshes during the wet

volcanic
sandstone

Chogo
eroded
pedotype

Chogo
ferrugin-
ized nodule
pedotype

hammer

Fig. 4.7 Two superimposed, brown, grassland paleosols (Chogo pedotype) 14 million years old interbedded with volcanic sandstones in the main fossil quarry at Fort Ternan National Monument, Kenya. The hammer handle extends down vertically from the label through the upper paleosol, which was buried by thick, volcanic sandstone. Both paleosols have dark crumb structure like those of soils in the area today, and imply comparable grassy vegetation 14 million years ago. The volcanic sandstones also preserve common fossil grasses

season. Other paleosols in the fossil quarries on Maboko Island are orange and studded with abundant hard, white, calcareous nodules at a shallow level within the profile, as in Aridisol soils of dry regions where rainfall is inadequate to leach calcium from the soil. These orange nodular paleosols are similar to soils of the African bushland called nyika, best preserved now in Tsavo National Park, Kenya. The global climatic cooling of the Monterey Event expressed itself in East Africa as a drier climate with more extensive grassland and bushland than before. A few weakly developed paleosols (Entisols) with the large root traces of woodland trees on Maboko yield a variety of fossil apes (*Mabokopithecus, Simiolus*) and bushbabies (*Komba*, a distant relative of living *Galago*). The nyika bushland paleosols contain a new kind of fossil ape, *Kenyapithecus afticanus*, which had short hands and feet, and wrists and ankles more like those of ground dwelling baboons and macaques than of monkeys habitually in the trees. *Kenyapithecus* is a likely common ancestor of apes and humans, and is also known from dry, grassy woodland (Mollisol) paleosols at the 14 million-year-old site of Fort Ternan (Fig. 4.7, Color Photo 4.3). The most common fossil in the dambo Mollisol paleosols

Color Photo 4.3 Late Miocene (14 million years) Chogo paleosols, Mollisols, with fossil grasses and antelope, in the Fort Ternan Member, Kericho Basalt, near Fort Ternan, Kenya

on Maboko Island is the oldest known African monkey, called *Victoriapithecus macinnesi*. This small monkey, no larger than a domestic cat, was ancestral to modern vervets, guenons, macaques and baboons. The presence of ancient grasslands at Maboko and Fort Ternan is confirmed by the abundant fossil antelopes and giraffes, which, like the monkeys, are the earliest known in this part of Africa. Remarkably though, early monkeys (*Victoriapithecus*) thrived in these grasslands, whereas our ape like ancestors (*Kenyapithecus*) retreated from these open habitats.

Once again grassland mosaic environments were associated with evolutionary divergence, this time between monkeys and apes. Mosaics of open and wooded grasslands, seasonally inundated grassland, thicket bushland and riparian woodland, offer a greater variety of habitats than tropical woodland alone. These varied habitats are apparent still in the national parks of Kenya: vervet monkeys clattering in streamside trees, dik-dik antelope ducking for cover in riparian shrubs, troops of baboons advancing through broken cover and herds of zebra and wildebeest in open grassy plains. Each species evolved within a particular portion of a mosaic of habitats.

Like the rise and fall of nations, major milestones in human and ape evolution have come at times of change for landscapes, when climate and vegetation changed in ways that were threatening to old life-styles. These changes are recorded in fossils and artifacts, but there is an additional record in the succession of buried soil horizons. Paleosols record changing vegetation and climate of the past. Red, blocky-structured Alfisol and Inceptisol paleosols record warm greenhouse climates of woodland and forest at 40,000, 130,000, 1.2 million, 3 million, 6 Million and 16 million years ago. Brown, crumb-structured, Mollisol paleosols record cooler and drier climates of mosaics of grassland and wooded grassland at 20,000, 150,000, 5 million and 15 million years go. Such mosaic habitats have encouraged evolutionary diversification, whereas more uniform habitats are associated with simplifications of ecosystems. These indications of evolutionary selection pressures from paleosols give a new perspective on human evolution as a product of alternating pruning and sprouting of our evolutionary tree. Mosaic icehouse environments encouraged diversification of our lineage, but more uniform greenhouse environments pruned back many of these new branches. From a Proserpinan perspective it is significant that expansion of carbon-rich grassland Mollisols were associated with cool climates and carbon-lean forested Alfisols and Inceptisols were associated with warm climates. The reasons for this will be explored in the next chapter, but probably did not include the rare and small creatures of African prehistory that we recognize from the increasingly detailed fossil record as our evolutionary ancestors. Cycles of global change on time scales ranging from thousands to millions of years have had profound consequences for our evolution. The current brisk pace of discovery and the fragmentary nature of fossils relevant to ape and human evolution are indications that there is much more to learn about our deep past from both fossil bones and soils.

HUMAN BEGINNING

The sun in Kenya is close and bright

And days pass quickly to starry night.

Hooves of wildebeest before the rains

Raise dusty columns on grassy plains,

Near lions basking, yawning in sun,

Guarding the carcass they have won.

Our parks and lawns recreate such lands.

Was this where we evolved our hands

Rising to walk erect, throw and display?

Old bones and soils tell another way.

Hands of humans first were good

To care for infants helpless in woods.

5

Grass that Changed the World

Sod grasslands and hypergrazers coevolved about 19 million years ago to create a novel kind of soil, Mollisols. Geographic expansion of these productive and carbon-rich soils cooled global climate.

Once on a mountaineering trip to Mt Aspiring in a remote part of the South Island of New Zealand, we were lucky enough to have 10 consecutive days of sunny weather, during which we climbed most of the peaks around the Bonar Glacier. This was an unexpected bonus because we anticipated no more than two or three days of climbing weather on such a long trip in this rainy region of New Zealand. But the biggest surprise after so long among rock and ice was the intoxicating, fetid stench of alpine grassland as we descended from the glacier to French Ridge Hut. The smell met us like a wall of thick moist air after the thin dry air of altitude. Our boots seemed to bounce on this living carpet after so long in rigid rock and ice. We spread ourselves on this dry, warm, living carpet for a picnic lunch, and looked down on the beech forests and meadows deep in the valley below. Later as we descended though cool and relatively odorless forest, with its hard gnarled branches and soggy uninviting soils, it seemed to me that there was something very special about grass.

On geological time scales, grasses, grasslands and their Mollisol soils are newcomers to our planet. The world was not always so well stocked with leaves of grass in alpine and polar meadows to tropical marshes and savanna. The oldest fossil grasses may be Cretaceous in age, some 70 million years old, yet grasses were not common during this age of dinosaurs. Remains of

© The Author(s), under exclusive license to Springer Nature
Switzerland AG 2022
G. J. Retallack, *Soil Grown Tall*,
https://doi.org/10.1007/978-3-030-88739-1_5

grassland soils indicate that their rise to prominence began some 30 million years ago, with sod grassland by some 19 million years and tall sod grassland by 5 million years ago. It is difficult to imagine a world without grasses, but outback Australia comes close. Driving west from the eucalypt forests of the mountains west of Sydney and into the flat riverine plains of inland New South Wales near Broken Hill is a very different experience than the drive west from Chicago to Denver in the North American Great Plains. In North America, groves of oak (*Quercus*) and hickory (*Carya*) forest in northern Illinois are largely gone by Iowa. Rolling open grasslands continue into sagebrush scrub by the Rocky Mountain front at Denver. In Australia the forests of blue gum (*Eucalyptus tereticornis*) and scribbly gum (*E. haemastoma*) give way to woodlands of white box (*E. alba*). Further west the trees become progressively smaller until shrublands of mallee (*E. viridis*) near Balranald and Mildura. By Broken Hill, red, rocky soils are speckled with scattered, low desert scrub of saltbush (*Atriplex*) and bluebush (*Maireana*). This vegetation pattern is broken by farms and pastures nurtured by two centuries of European settlement throughout inland Australia, but the steady decline in stature of largely woody plants toward the arid Australian outback is quite different to the wide-open prairie of North America. The Australian pattern is probably archaic, and is well documented from studies of 50–40 Million year old paleosols along paleoclimatic gradients. In other continents a whole new ecosystem of grasslands was interpolated between the forests of humid regions and the scrubby vegetation of deserts. This profound change from a world of arid-humid gradients in woody vegetation like those of outback Australia to a world with interpolated prairies had tremendous consequences for the global carbon cycle. As we shall see, it was probably not a coincidence that the evolution of grasslands over the past 30 million years coincided with long term planetary cooling and drying. Proserpinan mechanisms of global change can work on these long time spans as well as on the cycle of seasons.

In order to understand the prehistoric rise to prominence of grass and its role in long term global change, consider some reasons for the current success of grasses. There is more to grasslands than first meets the eye, much of it underground. Grassland soils or Mollisols are distinctive in the fine structure of their soil clods, which are small, rounded, dark, clayey and fertile (Fig. 5.1). Many of these small clods are fecal pellets of earthworms, which abound in grassland soils. The small clods also are sculpted by the close three-dimensional network of threadlike grass roots, which are very different in size and shape from the stout, woody, tapering roots of trees and shrubs. The clods of grassland soils are dark brown to black with finely comminuted organic matter intimately mixed with clay, unlike the brown to red plastic clays and

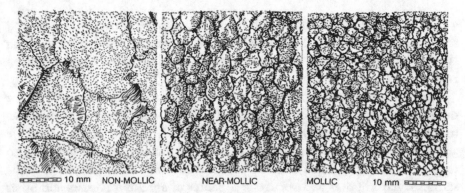

Fig. 5.1 Sketches of the distinctive crumb structure of soil clods in the soft, organic surface of sod-grassland soils (Mollisol: right), compared with the near-mollic clods of tussock grasslands (Inceptisol: center) and the blocky clods of forest soils (Alfisol: left), all drawn to the same scale

partly decayed twigs and leaves of woodland soils such as Alfisols and Ultisols. Grasslands have clays rich in plant nutrients such as calcium and potassium, whereas woodlands have clays leached of these elements. Clays of grasslands are stabilized by the films and stain of organic matter on and within the tiny ellipsoidal clods. The clays are thus restrained from shrinking and swelling into cracks and mounds as are found in the swelling clay soils or Vertisols, in which roots are broken, and on which water ponds in the wet season. Mollisols in contrast are remarkably fertile soils of humus and clay, stable after rain, well aerated, and teeming with life. Grasslands naturally create the tilth, or ease of working, that gardeners struggle to achieve with peat moss, lime, and worms. Many of these structural and root patterns are preserved in buried soils, which serve as a record of grasslands of the past.

Grasses are remarkable in their ability to colonize new surfaces disturbed by floods, landslides, or human construction. Their seeds are small, lightweight, and so insubstantial that they are blown across the landscape like dust. Once germinated and in flower, their pollen grains also are dispersed widely and in great quantities in the wind. The world is now blanketed with these tiny propagules so that seeds can germinate and grass flowers can be pollinated whenever conditions allow. Billions of seeds never germinate and many billions of grass pollen grains never find flowers. They are there nevertheless, waiting for the opportunity to make more grass until they dry and wither. This is a wasteful process compared with the carefully provisioned large seeds of a plum or peach. Such fruits have a large seed with much food for the germinating seedling, as well as a sweet outer flesh that encourages animals to eat the seed, carry it from the shade of its parent tree, and deposit it in warm, moist, fertilized ground. But then fruit trees are particular about good soils already littered with leaves and other plants through which the seedling is provisioned to grow, unlike grasses which germinate in open and disturbed land.

Some grasses form a tough green carpet, or turf, above the soil, and a durable underfelt, or sod, of root-fortified soil. Sod was once stacked in blocks to create sod houses of pioneers on the North American prairie. Now we are more likely to see it rolled up like carpet for ease of transplanting to create new lawns and golf courses. Turf and sod may have evolved to withstand the hard hooves and voracious appetites of antelopes, horses and other grazers. Grasses maintain these graze-resistant lawns with high growth rates. Some grasses grow more vigorously when grazed, the growth rate of leaves adjusting to the cropping rate. As a teenage boy given the chore of mowing our suburban lawn in Australia, my parents did not believe my theory that less frequent mowing might allow the grass to grow more slowly, but this idea has since found scientific confirmation in experimental study of Serengeti grasses and grazers. Grasses have the flexibility of growth rate to cope with heavy grazing in part because most grasses do not make much wood, which is a complex and slow process.

Grasses also have particular growth structures allowing them to withstand better than most plants the voracious appetites of grazing mammals. For example, the growing points of grass leaves are hidden beyond the grazer's bite (Fig. 5.2). The growing points or meristems in other plants are tender green areas of actively dividing cells at the ends of branches or leaves. In grasses, however, the meristems are hidden down among the basal part of the leaves that wrap around the stem. If you pull apart a grass plant, you can unfurl several leaf sheaths before coming down to a light green zone near the attachment to the stem. If you chew this tender zone near the base of the leaf, it gives sweeter juice and less grit than the outer part of the leaf. It produces new leaf in a way that pushes the frayed outer end of the leaf even further out. The ends of leaves are to some extent expendable, as long as their basal growth zones remain protected.

Not only have grasses' hidden growth mechanisms accommodated grazers, but grasses have also taken up arms against grazers in the form of small bodies of grit secreted in and on the leaf. These microscopic grit bodies are non-precious opal manufactured by the plant (Fig. 5.3). The mineral encrustation allows grass to stand dead and brown in the winter, after other herbs have wilted and rotted away. Hay is held together in part by this opaline silica encrustation. This silica grit in leaves of grass is like sand-paper to the teeth of grazing animals, such as horses and antelopes. The lives of many grazers in

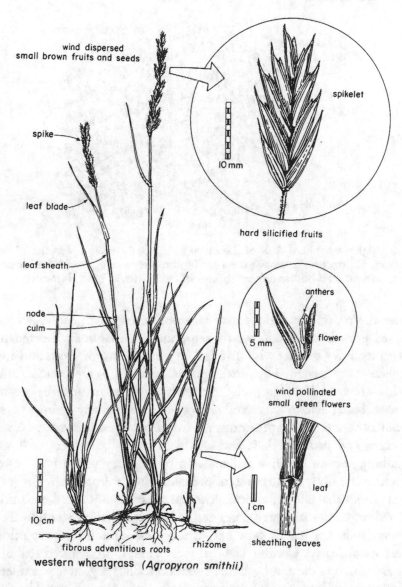

wind dispersed
small brown fruits and seeds

spike

leaf blade

leaf sheath

node

culm

10 cm

fibrous adventitious roots

rhizome

western wheatgrass *(Agropyron smithii)*

spikelet

10 mm

hard silicified fruits

anthers

5 mm

flower

wind pollinated
small green flowers

leaf

1 cm

sheathing leaves

Fig. 5.2 General features of a grass plant, as shown by western wheatgrass (*Agropyron smithii*) (reprinted from Retallack 2019, Soils of the Past, with permission of J. Wiley and Sons)

the wild are prematurely terminated by predators and disease, but otherwise they starve because of excessive wear on their molars. This is why grazing mammals have teeth with crowns that protrude high above the gum-line. These high-crowned, or hypsodont, teeth take longer to be worn back to the gums by abrasive grassy fodder.

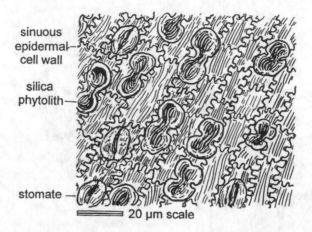

sinuous
epidermal—
cell wall

silica
phytolith—

stomate —

══════ 20 μm scale

Fig. 5.3 Highly magnified view of the outer cells of a 14-million-year-old (Miocene) fossil grass, *Cleistochloa kabuyis* from Fort Ternan, Kenya. This species has abundant dumbell-shaped opal bodies (phytoliths), as in living heavily grazed grasses

Over the past 50 million years, as grasses evolved a gritty silica crust, protected growth zones and weedy reproduction, mammals evolved high-crowned teeth for dealing with grit and elongate limbs for speed and spying over open grassy terrain. The grittier grasses survived to reproduce. Animals with low-crowned teeth ground their teeth to useless stumps before those with teeth better able to withstand abrasion. In open grassy plains where life was sometimes a race with predators, the future belonged to the fleet. Animals with claws and pads on their feet and with short limbs were well suited to climbing and hiding in woodlands. In open grassy vegetation, however, they were more likely to succumb to predators than animals with hooves and elongate limbs that allowed quick sprints to safety. Coordinated evolution of unrelated organisms such as grasses and grazers is called coevolution. Fossils and fossil soils can tell us how and when these coordinated evolutionary adaptations appeared. Commercial turf grasses and thoroughbred stallions of today are extremes created by human selection, but they merely accelerated evolutionary trends that have been operating for millions of years. The coevolution of grasses and grazers has changed many aspects of our world. Where did it begin, and how could we know?

For me the answer came like love at first sight. I was 26 years old and at the late spring beginning of a summer geological field camp. Our vans of geology students had just driven for two days over the vast, flat prairies of the North American Great Plains from Dekalb, Illinois, to an overlook in Badlands National Park, South Dakota (Fig. 5.4, Color Photo 5.1). My first view of these intricately eroded badlands not only took my breath away, but

Rockyford
Ash Member

rangeland
paleosols
of Brule
Formation

woodland
paleosols
of Chadron
Formation

Interior
pedotype

Yellow
Mounds
pedotype

Pierre
Shale

Fig. 5.4 A sequence of paleosols some 37–30 million years old in the Pinnacles area of Badlands National Park, South Dakota. The various named rock formations are characterized by different colors and textures of the paleosols within them. The two basal paleosols are named as soil types or pedotypes, and formed on marine shale of late Cretaceous age (68–70 million years old) and then river deposits of Eocene age (33–37 million years old)

made me eager for a deeper relationship with these beautiful rocks. I returned often over the next few weeks and decades because I suspected then, and subsequently confirmed, that the common brown paleosols lacking the abundant large root traces of woodlands and forests represented an early phase in the evolution of grasslands some 33 million years ago during the Oligocene epoch. These were the most fascinating paleosols that I had seen in North America, and there were so many of them, up and down the crenellated cliffs, and as far as the eye could see. I excitedly lectured the class on the fundamentals of the newly discovered root traces, soil horizons and soil structures, all diagnostic of paleosols. It is ironic that this landscape that captured the heart of a young geologist, and millions of modern visitors, should be called badlands. They are bad in the sense that they are difficult to traverse, and hard to cultivate. Lakota Sioux Indians called them "mako sica", later translated to "mauvaises terres" by French trappers, then badlands. Today they communicate wildness and raw natural forces in a world than has lost many of its former challenges. To me, they were an ideal combination of my twin loves of craggy, mountainous landscapes, and fossil remains of worlds past.

Color Photo 5.1 Late Eocene (37 million years) Interior paleosol (red Ultisol) in Chamberlain Pass Formation, under Gleska paleosols (Alfisols) in clayey Chadron formation, and Conata paleosols (brown Aridisols) in Brule Formation of Oligocene age (to 30 million years), Badlands National Park, South Dakota

The modern erosional features are as beautiful and inspiring as a Gothic cathedral, but it is the subtle color bands that give the badlands their charm. The slopes are streaked with red, dappled with pink and green, and high above the uppermost cliffs, smudged with brown. These colors tell a story of changing environments. Woodlands with their red and green Alfisol soils found in the lower and older rock layers of the badlands gave way later in time to brown Inceptisol soils of more open grassy vegetation. These earliest of grassland paleosols are brown with a fine structure of soil clods, but not such small, rounded clods as are found under modern turf grasses. Instead the clods are subangular and a quarter inch or so in size. They are also clumped in their distribution along the length of the ancient soil, as if the grasses grew in scattered clumps rather than as carpet-like turf. In addition to these features of tussock grassland soils, the paleosols also contain tiny fragments of plant opal with the sinuous, dumbbell, and rod form of opal produced by grasses. The fossil soils also have large, white calcite nodules, and they are shallow within the profiles, as in soils of dry regions. These paleosols dating back 33 million years to the Oligocene geological epoch are similar in many respects to soils now forming in dry intermontane basins of the western United States, where tussock grasses are interspersed with sagebrush and tumbleweed. Such vegetation can be called a dry rangeland or desert grassland.

The colorful fossil soils of Badlands National Park add much to their scenic charm, but the scientific fame of this region is rather for its fossil mammals, preserved in extraordinary quality and abundance. Many museums and rock shops have skeletons and skulls from here and from adjacent badlands extending many hundreds of miles from South Dakota into Nebraska and Wyoming. The appearance of open-country paleosols in these extensive badlands coincides with a dramatic extinction in fossil mammals about 33 million years ago. Large rhino-like creatures called titanotheres disappeared with the last of the pink and green paleosols of the Eocene epoch. New faunas of the Oligocene epoch were dominated by three-toed ancestors of horses and by extinct animals called oreodonts. These were sheeplike in general appearance, though with no indication of wool, and with short, padded feet more like those of dogs than sheep hooves. Oreodonts were abundant in some of the brown calcium-rich paleosols. Both oreodonts and early horses had a robust battery of grinding teeth adapted to coarse grassy fodder. Oreodonts and early horses also had elongate limbs, as in living mammals of open plains. Their teeth, however, were not so high crowned, nor their limbs so attenuated as in modern horses or antelopes. Coevolution of grasses and grazers had just begun at this time, and has been an ongoing process up to the present.

There was something missing in these earliest of grasslands. They did not form a sod. The bouncy alpine meadows of New Zealand were among my first introductions to the concept of sod, but living on the prairie soils of Illinois brought home this concept so unfamiliar to Australians. Grasslands in Australia have mainly bunch grasses along with small trees and shrubs, and much bare earth and woody litter is exposed. In Illinois however the ground is covered by a plush carpet of grasses and their roots binding a soil rich in humus. Grassland paleosols of the South Dakota badlands of 33 million years ago were more like the soils of Australian bunch grasslands. Sod-forming grasslands like those of the North American Great Plains today, did not appear until much later, in rocks of Miocene age some 19 million years old. The oldest known deep, brown, finely-rooted, crumb-structured paleosols are found at Agate Springs National Monument in northwestern Nebraska, another ragged line of badlands in the high open rangelands. There is little left in these paleosols of the organic matter that gives today's grassland soils their rich brown to black color. The organic matter of these Mollisol paleosols decayed away millions of years ago along with the fleshy parts of animals and plants that once lived on these soils. Nevertheless these ancient sods have abundant fine root traces, white with the mineral calcite, and their clayey paleosols are broken into crumblike soil clods. Concrete-like calcite not only fills root traces but forms nodules that are generally only 1–2 ft. (30–40 cm) below the ancient land surface. Such shallow nodular horizons are now found in soils of dry climates, receiving less than 16 in. (400 mm) rain per year, which is inadequate to leach calcite deep into the soil. There are other indications of dry climate also at Agate Springs. The paleosols at Agate Springs are famous for fossilized dens of bear dogs (*Daphoenodon*) and helical burrows of extinct relatives of beavers (*Palaeocastor*). These unusually deep burrows indicate both a very low water table and a need to retreat from a hot, dry land. These early sod grasslands appeared at a time of dry climate.

With these Mollisol paleosols of the earliest sod-forming grasslands of the Miocene appeared a new generation of fossil mammals. Three independent lineages of mammals, ancestors of horses, rhinos, and pronghorn antelope, all showed coordinated evolutionary change. Teeth in these fossil mammals were high crowned (hypsodont), projecting well above the gum line. The lower leg bones of horses and pronghorns, or "cannon bones", evolved from small hand and foot bones (metacarpals and metatarsals) of earlier mammals, which were more like dogs and humans in this respect. Thus in addition to the major limb bones (femur, tibia, humerus, ulna), horses and pronghorns have yet another robust limb element that elongated the limb considerably. The leggy look of racehorses that marks them as runners had appeared. Even so, few of these creatures were as elegant or toothy as a thoroughbred racehorse. The coevolution of grasses and grazers was accelerating, but by no means complete.

The hard hooves and large grinding teeth of these mammals are one reason why sod would evolve to keep the vital growing tips of grasses protected from the increasingly effective onslaught of grazers. But another reason has been suggested by Allan Savory of the Center for Holistic Management in Albuquerque, New Mexico, who has been a promoter of the old agricultural technique known as cell grazing. As an old Afrikaner proverb has it, "You have to hammer the veld to make it sweet". If cattle are kept in tight herds they eat both grass and unpalatable weeds to a close stubble. They defecate and urinate over the ground, and leave it unappetizing for several weeks, during which the most palatable, fast-growing grasses regrow in a dense sward. In contrast, cattle dispersed within fenced pastures graze selectively, creating spreading patches of woody and thorny weeds and deeply eroded trails to water, with the result that pasture quality declines. Generations of shepherds and cowboys have been good for the land, but in nature it is the large cooperative predators such as lions and wolves that keep grazers within tight herds. Intriguingly, the appearance of sod grassland paleosols in North America coincides in time with a number of indications of pack hunting and herding behavior. The prorean gyrus is a lobe in the brain of mammals clearly visible from casts of the inside of the skull, and it first appears in bear dogs (*Tomarctus*) of North America about 19 million years ago. Before that time, the canine and feline predators of North America were more likely solitary hunters. By 19 million years ago we have the first clear evidence of herds, in 17 articulated skeletons of camels overwhelmed by a dust storm in the *Stenomylus* Quarry of the Carnegie Museum near Agate Springs. Among the millions of earlier Oligocene fossil oreodonts, I know of no more than 10 articulated skeletons preserved in a group. These are parents and a litter of pups fossilized within a den, now at the Tate Museum in Casper, Wyoming. I envisage oreodonts like a family group of kangaroos in the Australian outback, but not like tight and vigilant herds of impala or wildebeest on the African plains. Finally, the earliest evidence of cake-like dung comes from pancake turds preserved with 19 million years old grassland paleosols at Eagle Crags near Agate Springs in Nebraska. I have only found fossil pellet-like turds, like those of kangaroos, as evidence for the defecation of Oligocene oreodonts and other mammals. Herding induced by predation goes back at least 19 million years, and as Allan Savory has indicated, was vital for maintaining rangeland productivity.

Sod grasslands of 19 million years ago were still not quite modern in that they were not yet widespread, extending no further east in North America than Nebraska and South Dakota where the nodules of calcium carbonate were no deeper within the soil than two feet (50 cm). Sod grasslands had not yet rolled out over the landscape into tall grassland soils more deeply leached of calcium carbonate like those of humid Illinois and Indiana. Tall sod grassland soils did not appear until 5–7 million years ago, and are represented by paleosols of that age in gullies and low cliffs of the high plains of western Kansas, near the towns of Ellis and Hays. The paleosols there also have small granules of brown clay and abundant fine root traces of Mollisol soils formed under sod-forming grasslands. In these paleosols, however, the calcite nodules are as deep as 3 ft. (1 m) down in the profile and the brown granular layer is thick and even textured. The observation that calcite nodules are deeper in soils of humid climates and more productive vegetation, and shallower in soils of dry, less-productive vegetation is widely used to determine land use by rangeland managers. For the paleosols near Ellis, this rule of thumb indicates that grasslands had extended their climatic range into wetter regions receiving perhaps 30 in. (750 mm) of rain per year, near the current rainfall zone where grassland passes into woodland. These 7-million year old Mollisol paleosols are the oldest known that are likely to have supported tall-grass prairie like that now growing near Lincoln, Nebraska.

Coincident with the appearance of tall grassland paleosols in late Miocene rocks of 5–7 million years ago, the fossil record of grasses improves substantially. A few fossil grasses and many tiny pieces of grass opal have been found in older rocks of Kansas, Nebraska and South Dakota, but in late Miocene rocks there is surprising diversity and abundance of fossil grass husks and leaves. This diversification of fossil grasses does not entirely reflect a late Miocene evolution of grasses, because many distantly related kinds of grasses are found. What changed was preservation of a greater variety of grasses. Most of the fossils are flower husks encrusted with opaline silica produced by the plant. These opal-encrusted parts of the plant were preserved after the organic parts of the plant decayed, similar to the way in which the soft parts of snails decay, leaving behind their shells. A variety of different kinds of grasses had evolved the gritty, silica defense against a formidable array of grazers with high crowned teeth and hard hooves. The single-toed horses and cloven-hooved American antelope of this time were already close to modern in appearance. Coevolution of grasses and grazers was now in full swing.

A remarkable feature of the advent of tall grasslands some 7 million years ago is that it was synchronous in most parts of the world, whereas earlier stages in the evolution of grasslands were not. Short sod grasslands are represented by paleosols as old 19 million years in North America, but only 15 million years in East Africa. Tall grasslands, on the other hand, appeared some 7 million years ago in North America, South America, East Africa, Pakistan, and Australia. Reasons for this synchroneity have recently become apparent from studies of isotopes in fossil soils and teeth.

Isotopes are different weights of the same chemical element. Carbon is an element with one unstable or radioactive isotope, ^{14}C, used for radiocarbon dating, and two stable isotopes, ^{13}C with 6 nuclear protons and 7 neutrons, and the lighter ^{12}C with only 6 of each. Most of the carbon in the world is ^{12}C, but ^{13}C and ^{14}C are rare. All the detectable ^{14}C is decayed to ^{12}C after about 70,000 years, and so not measurable in carbon within rocks older than that. The other two isotopes are geologically stable and preserved in paleosols as a record of changing vegetation. Grasses are important to the ratio of the stable isotopes in nature because they use at least two kinds of photosynthetic processes that affect isotopic composition of carbon. Many grasses are like most other plants in using Calvin-Benson cycle photosynthesis that produces a 3-carbon phosphoglyceric acid as a basis for growth, and so is called C_3 photosynthesis. Enzymes active in this metabolic process select strongly for ^{12}C rather than ^{13}C in carbon dioxide (CO_2). Calvin cycle plants thus produce organic matter with isotopically lightweight carbon. This organic matter decays in the soil to make isotopically light carbon dioxide, which is incorporated into isotopically light calcite ($CaCO_3$) of soil nodules.

A second photosynthetic pathway found mainly in tropical to warm temperate lowland grasses is the Hatch-Slack or C_4 pathway, so named for the 4-carbon malic and aspartic acids produced as a starting point for subsequent plant metabolism. This pathway includes a gas-recycling process in special cells around veins of the leaf, and is less discriminatory against the heavy isotope ^{13}C. As a result the soil organic matter, carbon dioxide and nodule calcite under C_4 grasses are isotopically heavier compared with those under C_3 grasses by a significant amount of 5–10 parts per thousand on a scale of measurement pegged to an arbitrary standard of Cretaceous marine carbonate values. Comparable isotopic differences are also transferred to the teeth of mammals eating such grasses. Thus the ratio of carbon isotopes in organic matter, in calcite nodules of paleosols, and in the apatite of mammal teeth can provide useful information on diet and vegetation of the past.

Both paleosol nodules and mammal teeth in such disparate parts of the world as Nebraska, Bolivia, Argentina, Kenya, and Pakistan, became isotopically heavier, signifying the appearance of tall C_4 grasslands in tropical and subtropical regions some 7 million years ago. Distinct regional fossil grass floras indicate that C_4 grasses did not migrate to such widely scattered regions. One could argue that C_4 grasses, which are now mainly tropical, spread because the world became hotter, more seasonal or flatter at this time, but this is not borne out by the varied monsoonal and warm temperate climatic regions in which the isotopic shift is found. A globally unifying explanation comes from the observation that C_4 grasses are more successful than C_3 grasses under conditions of low atmospheric carbon dioxide. An atmospheric carbon dioxide decline beginning 7 million years ago was already suspected for other reasons. Fossil leaves of oak (*Quercus petraea*) and ginkgo (*Ginkgo biloba*) show increased abundance of the microscopic breathing pores or stomates at about 7 million years ago. Experiments with living relatives of these fossils demonstrate that declining carbon dioxide would elicit this anatomical change. Carbon dioxide from the burning of fossil fuels is a major contributor to current global warming by the greenhouse effect, because carbon dioxide is a large molecule that traps the sun's warmth like the glass panes of a greenhouse. Conversely, periods of low carbon dioxide are times of global cooling, glacier growth, and sea-level drop. At 7 million years ago there is geological evidence of major advances of the Antarctic ice cap, and of sea level falling so low that the Mediterranean sea dried up to become a parched salt desert. Tall grassland thus appeared at a time of cooling, drying and global atmospheric decline in carbon dioxide concentration.

Such coincidences in time invite speculation about cause and effect. Did grasses change their world or did they benefit from a world changed by other causes? The idea that grasses exploited cooler and drier climatic shifts has been popular in recent years, with the prime suspect for cooling the globe being uplift of Himalayan Mountains. The argument is that the inexorable movement of India into Asia pushed up this giant among mountain chains, initiating glaciers and intensifying the South Asian monsoon. It is an appealingly simple idea, but one that runs into serious difficulties. Principal among these is that frigid soils of the Tibetan Plateau would have consumed less atmospheric carbon dioxide during weathering than the warm and lush ecosystems they are presumed to have replaced, and so are unlikely agents of global cooling. Mountain uplift curtails chemical and biological weathering at cool temperatures of altitude, encourages glacial growth, and sheds glacial debris and other sediments at rates faster than can be weathered by lowland ecosystems. How could mountain uplift promote the biological consumption of carbon dioxide they were working to curtail?

A second idea is that accumulation of salts in the land-locked basin of the Mediterranean curtailed production of dense, cold, saline, deep water in the North Atlantic. These cold saline waters are part of a global oceanic circulation system, exchanging cold deep polar water for warm shallow tropical water. Disruption of that circulation could produce cooling and drying. However, the northward tectonic drift of continents was opening rather than restricting the global heating system of ocean currents. As Australia, Africa and South America drifted further north from Antarctica, the southern oceans widened, allowing more water to flow through the great global heat pump of ocean currents. North Atlantic cold water now sinks and passes south at depth into the Southern Ocean, upwelling off the coast of Chile, then warming as it filters through Indonesia and the Indian Ocean and on into the warm surface Gulf Stream of the North Atlantic. This pattern of circulation has persisted since at least 37 million years ago, but amplification of these oceanic currents in the progressively wider southern oceans created by continental drift would have warmed, not cooled, the planet. The most important constriction of thermohaline circulation was the expansion of sea ice around Antarctica some 5-million years ago. How could the ocean currents have created the sea ice they were working to prevent?

An alternative view is that newly evolved tall grasslands represented a different and now widespread ecosystem that altered the carbon cycle and changed our world. Unlike physical causes of climatic cooling, these mechanisms are ultimately biological. The grasses are mindful of grazers and the grazers of grasses, and neither is concerned with climate change, just as humans follow their livelihoods regardless of consequences. Rather than grasslands expanding as more regions became cooler and drier, grasslands advanced into wetter and more productive soils with deeper nodules of calcite as a result of the continued coevolution of grasses and grazers (Fig. 5.5). Harder-hooved and higher-toothed mammals presented an onslaught on vegetation that few plants other than grasses could resist. Bulldozer herbivores, such as rhinos and elephants, also were very destructive of trees. After fires, grasses regenerate more readily than trees and shrubs, and annual grasses are especially flammable during dry and cold seasons. Fire, overgrazing by horses and antelope, and tree demolition by rhinos and elephants were a new ecological regime that advanced the empire grass to a new equilibrium with woodland.

These coevolutionary developments had important consequences for the carbon cycle. Grassland soils (Mollisols) are dark with organic matter and much richer in carbon than the dry woodland and desert scrub soils that they replaced (Alfisols and Aridisols). Although woodland and desert soils (Alfisols and Aridisols) can have much humus near the surface (4 in. or so), many tall grassland soils (Mollisols) are dark brown with organic matter to depths of 3 ft. (1 m) or more. Tall grasslands of warm regions sequester more carbon than any other soil type other than peats (Histosols), which cover only 3% of land area. When broad expanses of the North American Great Plains, the Russian steppe, the Argentine pampas, the Serengeti Plains and northern Australia were covered with tall grasses, enormous amounts of carbon were taken from the atmosphere and stored temporarily in soil, and then permanently stored in organic soil clods eroded from soils and deposited in lakes, rivers and the sea. Grassland soils and sediments, by stealing carbon dioxide from the atmosphere may have diminished the earlier greenhouse climate and brought on climatic cooling of the Ice Ages.

Grassland soils also influence climate by more efficiently retaining water than soils of woodlands and deserts. Trees and woody shrubs are active in extracting water with their roots and transpiring it through their leaves, creating moist air, but dry soil. The grassland sod in contrast traps moisture within the soil. Water vapor is another greenhouse gas like carbon dioxide, so the cutting of woodland for pasture has the effect of encouraging a drier and cooler climate. Moist soil of grasslands with small crumb clods presents a higher internal surface area for weathering than desert or woodland soils, and this accelerates weathering by dilute carbonic acid as a more effective carbon sink in grassland soils.

Finally, grasslands are lighter colored than woodlands, especially with winter dieback, and so reflect more light and heat back into space. When

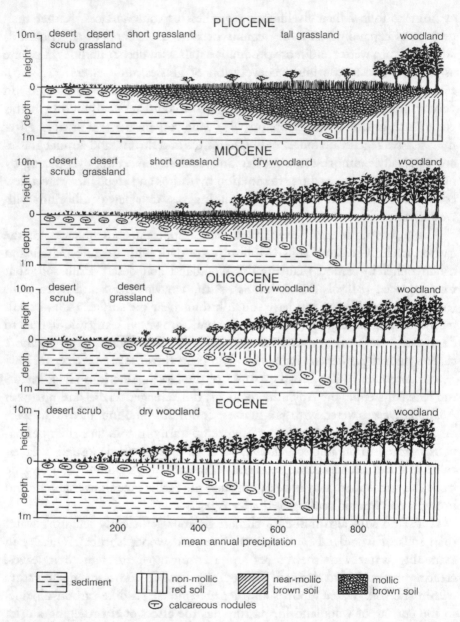

Fig. 5.5 Three-stage evolution of grasslands on a climatic transect in the Great Plains of North America through the past 50 million years, including Oligocene desert grasslands beginning at about 33 million years ago, Miocene short-grass prairie beginning about 19 million years ago and Pliocene tall-grass prairie beginning about 7 million years ago (reprinted from the journal Palaios, 1997, with permission of the Society for Sedimentary Geology)

snow falls on a grassland it creates a highly reflective surface that can cause snow-blindness on a sunny day. Snow on a woodland in contrast is more readily shed to reveal less reflective branches and leaves. Thus, in addition to their impressive carbon storage, grassland soils cool and dry our world by curbing atmospheric water vapor, consuming carbonic acid, and reflecting solar energy.

Paleoclimate over the past 50 million years has changed from a warm, humid greenhouse, to a cold, dry, ice age in a sequence of coolings at 33, 15, 7, 2.5, and 1.7 million years. Grasslands expanded their climatic and geographic range, and showed evolutionary innovations in size, teeth, hooves and plant opal at each of these times. There have always been mountains and ocean currents, and a complex system of interactions between organisms and their environment, but grasses are the principal new element of the past 50 million years. These novel carbon-hungry ecosystems replaced dry woodland and desert scrub in the current grassland belts of the world receiving some 12–40 in. (300–1000 mm) of rain per year. In historic times, grasslands have expanded even further beyond this climatic range because of human forest clearance, and pasture improvement. This may have saved us from excesses of global warming due to forest and fossil fuel burning. Small evolutionary improvements in grasses and grassland ecosystems have had profound consequences. On such long time scales, the Proserpina Principle operates through ecosystem evolution rather than ecosystem development. In this history of grasslands as revealed by their fossil soils, plants and animals, biology was destiny.

There is also a message here for the future. Restoring grasslands could be helpful in sequestering atmospheric carbon dioxide of our current greenhouse crisis. Just as carbon storage in expanding grassland soils ushered in a global ice age, carbon storage and water capacity in soils can be improved by agricultural practices, such as cell grazing used with success by Alan Savory and Richard Teague, among others. Cell grazing, contour coppicing, and pasture cropping are but three of many techniques available to carbon farming, which is already central to "organic" vegetable production. With subsidies to farmers shown to build carbon in their soils coming from government carbon taxes, carbon can be taken from the atmosphere and returned to soil. Grasslands once covered the most productive soils on Earth, not so arid as to be limited by water, and not so humid as to be leached of essential nutrients. These same productive soils are almost all under the direct management of farmers. Just as these soils and their herds once brought on the Ice Ages, they could be used again to draw down greenhouse gases.

SUPERSOD

The form of woman or up-reaching tree

Is praised in song by all who see,

But who speaks for the glory of grass,

For leaves of green, for hay like brass,

For brown earth's sharp smell that lingers,

For small round clods that fall through fingers

Without smear, for deep and pliant sod

That softens where weary feet have trod.

The teeming soil takes festering dead

Not sickening itself, sending up instead

Slender green shoots of timid grain,

Turning over the cycle of life again.

6

Death from the Sky

The mass extinction of 66 million years ago that extinguished forever favorite fossil lineages such as dinosaurs and ammonites, also left a record in paleosols. Impact beds from a large bolide impact disrupted soil formation and vegetation in Montana.

The ships of the invading Greeks had left. Their seaside camp was abandoned. The Trojans, stunned and surprised at first, celebrated their victory in the long war by dragging into the city the large wooden horse left by the Greeks. In the dead of night, Greeks emerged from the horse and throwing open the city gates to fellow warriors, destroyed the city and its people with sword and flame. Archeological excavations at Troy show that this was one of 13 sackings of the prosperous spice-trading and horse-rearing community in what is now northwestern Turkey. Disaster can strike when you least expect, undoing the efforts of generations.

The successive layers of buried cities of ancient Troy, Mohenjo-Daro, or Ur, are a record of destruction and rebuilding, like a succession of fossil soils, one on top of another. Each of the pink and brown paleosol layers in Badlands National Park of South Dakota represents a time of tranquility and soil building. The layers of sand and volcanic ash, on the other hand, represent volcanic eruptions and floods that destroyed existing local order. These records of alternating soil formation and disaster show great variation in both duration of soil formation and magnitude of the disaster. Even thin layers of volcanic ash or flood-borne silt can beat down and bury streamside herbaceous vegetation rooted in weakly developed soils. Such small disasters barely register in forest soils. Worms and roots quickly mix thin sediment layers into pre-existing soil. It takes a profound disaster to bury a forest and

© The Author(s), under exclusive license to Springer Nature
Switzerland AG 2022
G. J. Retallack, *Soil Grown Tall*,
https://doi.org/10.1007/978-3-030-88739-1_6

its thick clayey paleosol. We know from experience that the biggest floods and volcanic eruptions come at long intervals. The 1000-year flood is much more destructive than the 100-year flood used for flood insurance estimates. The 29-million-year-old Rockyford Volcanic Ash in the badlands of South Dakota, Nebraska, Wyoming and Colorado, is a record of a volcanic eruption many times the size of any known historic eruption. There were similarly massive eruptions of hot glowing ash of comparable geological age in the high desert badlands of central Oregon. Fortunately, these eruptions that lay waste to thousands of square miles are rare events. In Oregon there were only a dozen of them over the last 40 million years. Their effects on local soils were evidently transient, because similar paleosols are found below and above these thick volcanic ashes. Despite total devastation of large areas covered by tuffaceous rocks, colonizing animals and plants from surrounding areas rebuilt woodlands and sagebrush communities similar to those before disaster struck.

But there are disasters facing life and soils even greater than voluminous clouds of volcanic ash. Large asteroid and comet impacts with the earth are also rare and come at long intervals. There is a continual rain of cosmic dust onto soils and into the seas of the Earth, but larger stones come at longer intervals. Most burn up as shooting stars, but some make it to the ground before consumption in flame. Michelle Knapp of 207 Wells St, Peekskill, New York, was surprised at 7.50 p.m. on October 9, 1992, when a 29-pound stony meteorite smashed the trunk of her Chevrolet Malibu parked in the driveway. Such cosmic gifts are rare, but bigger rocks from space are even rarer. A meteorite some 150–300 ft. in diameter made the Barringer Meteor Crater in Arizona some 49,000 years ago. This stark circular bowl in the desert is 4100 ft. (1250 m) in diameter and 600 ft. (183 m) deep, its rim rising to 200 ft. (61 m) above the surrounding plain. Rock and soil from the crater was blasted for miles around, covering soils and initiating a new round of soil formation. Asteroids miles across fall at intervals of tens of millions of years. Those tens of miles in size come at intervals of hundreds of millions of years, and have profound effects on life and soil.

Large asteroid impacts are devastating not only near the impact crater, but for much of the planet. Computer simulations have revealed a variety of catastrophic effects, comparable to those of a full-scale exchange of nuclear weapons between the United States and Russia. The kinetic energy of a big impact is transformed on impact into an enormous fireball that vaporizes rock, soil, and forests for hundreds of miles around. Heating and compression of atmospheric gases in the path of the impactor, and vaporization of salt deposits at the point of impact, create the noxious gases nitrogen dioxide and

coal-bearing
paleosols of
Tullock
Formation

Cretaceous-
Tertiary
boundary

gray clayey
paleosols of
Hell Creek
Formation

Fig. 6.1 The boundary between the Cretaceous and Tertiary geological periods, some 66 million years ago, exposed in a sequence of paleosols in the badlands of Bug Creek, eastern Montana, U.S.A.

sulfur dioxide, which dissolve in rainwater to form strong acids that chemically scald trees and leach soils. A cloud of melt-glass and rock fragments is thrown up by the impact. Dust and debris are lifted high into the atmosphere, cutting off the sun, and initiating torrential rain and snow. When the dust settles, carbon dioxide from vaporization and burning of plants and soils warms the atmosphere and continues the leaching of soils with carbonic acid. Such dire simulations can seem as divorced from reality as prophecies that the end of the world is near, but prehistoric records reveal awful details of such past disasters.

One place where such a violent cosmic catastrophe is becoming well known is in the sequence of paleosols and river deposits, 60–70 million-years-old, in Hell Creek, eastern Montana (Fig. 6.1, Color Photo 6.1). This dry grassland with its barren badland slopes is well named. My six-year-old son Jeremy was as eager as many children his age to find a banana-sized tooth of that great dinosaur *Tyrannosaurus rex* on our visit there. His enthusiasm flagged rapidly in simmering summer heat of 102 °F (39 °C). If the heat does not get to you, then it is fierce wind and rain from summer thunderstorms. Winter is bitterly cold and snowy, but still fossil hunters comb the badlands for the big bones of the dinosaur king *T. rex*, his three-horned prey *Triceratops horridus*, and a glorious diversity of other dinosaurs. Hell Creek preserves a record of the last dinosaurs, as well as the small mammals that inherited the Earth after dinosaur extinction. Few areas in the world have such a complete

Color Photo 6.1 Cretaceous-Tertiary boundary (66 million years) between Ottsko paleosols, gleyed Alfisols, and Sikahk paleosol, gleyed Ultisol, of Late Cretaceous Hell Creek Formation, and Sik paleosol, Histosols, of the Paleocene Tullock Formation in Bug Creek, Montana

record of this extinction estimated to have destroyed some 60% of species on land and in the sea.

In a low hill called Brownie Butte near Hell Creek are inconspicuous layers of clay, about 66 million years old, between the last of the dinosaurs and the first of the surviving mammals. These clay beds were the real reason for my visit with Jeremy. The thin claystone beds run through a black peaty layer that once formed a swampy Histosol soil like those now forming in Georgia's Okefenokee Swamp. One of the claystone layers is a little less than an inch thick, and light pink in color. This layer has been called the boundary bed, because it is a marker for the boundary between the Cretaceous and Tertiary periods of geological time (Fig. 6.2, Color Photo 6.2). Immediately above it is a layer of gray shale less than a half inch thick that represents an extraordinary event. The gray band is rich in iridium, a very rare element at the Earth's surface. Claystones anomalously rich in iridium were initially discovered within a gully north of Gubbio in the Italian Alps, which was at the floor of a deep ocean at the time of the Cretaceous-Tertiary boundary. Later it was found that iridium was enriched at the same time on land as well.

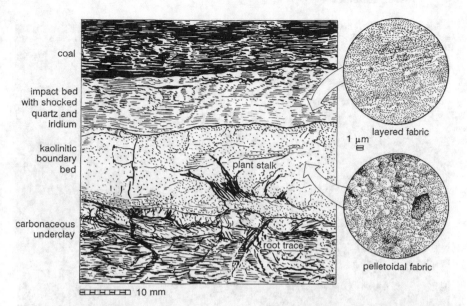

coal

impact bed
with shocked
quartz and
iridium

kaolinitic
boundary
bed

carbonaceous
underclay

plant stalk

root trace

layered fabric

1 μm

pelletoidal fabric

10 mm

Fig. 6.2 Close-up view of the thin boundary beds that mark the Cretaceous-Tertiary boundary at Brownie Butte, near Hell Creek, Montana. Comparable and distinct boundary and impact beds have been found at numerous localities in the North American west, and are regarded as fallout from a large asteroid impact in Yucatan, Mexico

Subsequently, the iridium anomaly was found in more than 100 latest Cretaceous localities worldwide, including Brownie Butte. These discoveries were initiated at the University of California in Berkeley by a team of scientists, headed by the Nobel-prize winning physicist Luis Alvarez, and his son Walter Alvarez. The iridium provided stunning evidence that linked the extinction of dinosaurs and other late Cretaceous creatures to the impact of an asteroid or comet. Iridium is a rare element in most rocks on Earth, but is abundant in meteorites. There were few plausible alternatives to the astounding idea that iridium enrichment at the Cretaceous-Tertiary boundary came from a large asteroid or comet vaporized and globally distributed on impact with the Earth. From the amount of iridium found at the Cretaceous-Tertiary boundary and known in likely meteorites, Alvarez and his team were able to make a rough calculation of the size of the impacting body. A mountain of rock 6 miles (10 km) in diameter, its impact had globally catastrophic effects.

The iridium anomaly and its interpretation as evidence of end-Cretaceous catastrophe stimulated a storm of controversy among scientists. It is unsettling to think that a rock from space could nearly destroy our world. Perhaps the iridium came from volcanoes, rather than from an impacting asteroid? There was a large eruption at the time of the Cretaceous-Tertiary boundary,

Color Photo 6.2 Cretaceous-Tertiary boundary (66 million years) at millimeter scale (below) between last Sikahk paleosol of Cretaceous (below) with frayed shoot sticking up into kaolinitic impact bed, and then smectite-jarosite impact bed, followed by coal of Sikahk paleosol in Hell Creek Formation, near Brownie Butte, Montana

but these basalts called the Deccan Traps of India were quiet flows, not explosive eruptions that loft dust high into the atmosphere. This alternative volcanic hypothesis was soon buried by an avalanche of supporting evidence for asteroid impact. The thin gray claystone at Brownie Butte, as at many other locations with the iridium anomaly, also yielded quartz grains criss-crossed with glassy planes of a kind produced only by the titanic shock forces induced by meteorite impact. Shocked quartz created within and around the impact zone was blasted out with the iridium-rich dust and other ejecta. The gray claystone at Brownie Butte also contains enigmatic spherical grains. Similar tiny spheres have been found in Cretaceous-Boundary beds in Mexico and around the Caribbean region, where they are better preserved. They are glass spheres from the quenching of droplets of impact-melted rock. In Texas, Alabama, and Mexico, the Cretaceous-Tertiary boundary is marked by thick sandstone beds that can be interpreted as deposits of giant tsunamis generated by impact in the Caribbean area. Finally, a giant crater some 110 miles (177 km) in diameter was found shallowly buried in the Yucatan Peninsula of Mexico. Melted rock recovered by drilling into this impact structure has been dated using radioactive isotopes, and is the same geological age as the pink and gray claystones marking the Cretaceous-Tertiary boundary in Montana and elsewhere. So many lines of evidence indicate impact that few scientists dispute an asteroid or comet impact at the Cretaceous-Tertiary boundary, although argument continues as to how this affected life and landscapes.

The Montana badlands of Hell and Bug Creeks are a detailed record of impact and its consequences. At levels in the sequence yielding dinosaurs as well as in higher levels formed when they were gone, fossil soils are gray and green in color, similar to Inceptisol soils of swampy lowlands (Fig. 6.3). Peaty Histosol paleosols are found both above and below the pink-gray boundary claystones from the impact debris. This contradicts the idea that profound climate or landscape changes caused the extinctions. Dinosaurs weathered many comparable climatic shifts in their 130-million-year evolutionary career before demise at the end of the Cretaceous. Similarly, paleosols reveal no great disruptions in water tables or landscapes that could be blamed for loss of the dinosaurs. There are many thin volcanic ash layers both in the beds with dinosaurs and in those that accumulated after their extinction. Even in India, the first enormous basalt flows of the Deccan Traps were survived by dinosaurs who made extensive nesting grounds in paleosols preserved between the flows.

A rather different story comes from thin bands of claystone in the coaly parts of the paleosol at Brownie Butte, Montana. Curiously, all the iridium, shocked quartz and spherical melt droplets from asteroid impact are in the thin gray shale. The underlying pink claystone does not contain these indications of impact, but instead was produced by lethal effects of the impact. The gray-pink claystone couplet is found not only at Brownie Butte, but at some 30 sites in North America from Saskatchewan south to New Mexico. The pink claystone has been strongly leached of calcium, magnesium, sodium, and potassium compared with the gray claystone layer above. Such a distinctive alumina-rich layer could be formed by weathering over tens of thousands of years. But in this case, the pink claystone contains trace elements indicating that its source was identical to that of the overlying gray shale. The pink claystone is altered debris from the impact which fell only hours before the dust settled completely to create the gray shale. The distinctive composition of the thin pink claystone can be explained if it consumed much more acid than the gray shale. Acid leaching may also explain the iridium-poor composition of the boundary bed.

Acid has long been suspected as a product of asteroid impact. The passage of an asteroid through the atmosphere at high velocity can shock nitrogen in the air to noxious oxidized nitrogen, which later combines with rain to form nitric acid. But the Caribbean impact of 66 million years ago really

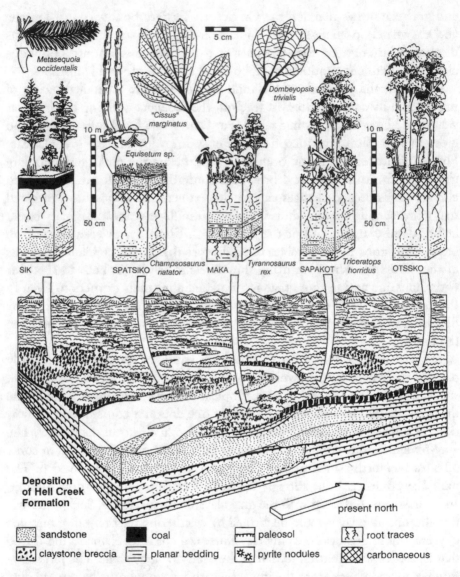

Fig. 6.3 Reconstruction of soils, plants and animals of Bug Creek, eastern Montana, during latest Cretaceous time, some 66 million years ago (reprinted from the Bulletin of the Geological Society of America, 1994, with permission from the Geological Society of America)

hit the powder keg, by vaporizing thick deposits of sulfur salts to sulfur dioxide gas. The 120-mile-diameter Chicxulub Crater in the Mexican state of Yucatan blasted sedimentary rocks that included thick deposits of the sulfur salt gypsum. These salt deposits were formed millions of years before by evaporation of locally restricted ocean basins, and are found in only a few small parts of the world. This was very bad luck indeed, because an already enormous amount of nitric acid may have been supplemented by large amounts of sulfuric acid. Much of this acid would have been consumed during a vast acid–base reaction in midair as melted rock thrown up by the impact cooled to a point of water condensation allowing acid attack. This initial neutralization of acid by reaction with rocky debris probably created the pink claystone. This was fortunate because it blunted the assault of these noxious gases and acids as they were spread over much of the globe.

For the dinosaurs this choking and scalding mix would have been lethal. But the soil itself remained as a last defense for smaller creatures, soaking up acid and neutralizing it by alteration of minerals. The buffering capacity of the calcareous and clayey soils of Montana can be calculated from the chemical composition of the paleosols. Each calcium and magnesium ion is displaced by two hydronium ions (essentially protons) of acid, and each sodium and potassium by one hydronium ion. The buffering capacity of the soils was sufficient to keep groundwater pH above the potentially sterilizing value of 4, so that many fish and amphibians survived the catastrophe. Clams and snails on the other hand disappeared as pH declined below a value of 5.5 required for them to form shells. Among the rat-sized mammals of the time, those that ate insects fared better than those that required fresh meat, leaves, or fruit. Many broadleaf evergreen flowering plants also became extinct as their leaves were browned by acid. Surviving plants were mainly deciduous broad-leaved flowering plants and needle-leaf conifers which regenerated from the soil's seed bank. The full brunt of acid generated by such a large impact could have been devastating, but its buffering to manageable levels by Montanan soils made the crisis a selective gatekeeper into the Tertiary. In granitic and quartz-rich soils of other regions already acidic in reaction, the gateway into the Tertiary may have been closed completely.

Acid rain was only the beginning of an unfolding disaster for global climate. Dust thrown up by the impact together with soot from burning forests would have created global darkness and chilling. Such conditions would have been comparable to the "nuclear winter" predicted from full-scale nuclear war. The dust settled after a period of months to years, and the warmth of the summer sun could be appreciated again.

This new dawning was further warmed by residual atmospheric carbon dioxide from global death and decay. To this can be added carbon dioxide from thermogenic cracking of coal and other organic matter by the gigantic feeder dikes to the large basaltic eruptions of the Deccan Traps of peninsular India, which cluster in geological age around 66 million years ago. Transient greenhouse warming is revealed by the changing isotopic composition of plant debris and coals at the Cretaceous-Tertiary boundary. A short-lived (about 50,000 year-long) shift to carbon compositions richer in ^{12}C and more depleted in ^{13}C than before, indicates a massive oxidation of plant biomass to carbon dioxide, or injection of isotopically light methane that would oxidize to carbon dioxide. More evidence comes from small breathing pores (stomates), which are sparse in fossil *Ginkgo* leaves found in the shales just above the Cretaceous-Tertiary boundary, compared with more common stomates in *Ginkgo* fossils below and well above the boundary beds. Plant leaves have fewer stomates when their fuel of carbon dioxide is abundant, and so indicate a short-lived carbon-dioxide greenhouse after the Cretaceous-Tertiary boundary. There also are clues from paleosols less rich in calcium carbonate above the boundary, than paleosols below. Less carbonate is expected in soils of wetter and warmer climates, with more heat and rainfall to leach calcium from soil minerals. Such comparisons suggest a paleoclimatic shift from some 35–47 in. (900–1200 mm) mean annual rainfall in the latest Cretaceous to 47–59 in. (1200–1500 mm) in the earliest Tertiary. Histosol paleosols are much more common in sediments formed after the impact in Montana, but in nearby South Dakota, which was closer to the sea to the east and thus more swampy, these peaty paleosols are common both above and below the boundary. Thermal expansion of the oceans and melting of montane glaciers at this time of geologically transient warming may also explain global sea level rise and westward spread of swamps into Montana after the great extinction.

Acid rain and global warming are pressing issues today, because the burning of fossil fuels releases acid-generating sulfurous and nitrous gases, as well as carbon dioxide that forms carbonic acid in water. Life will bounce back from our insults, just as it bounced back from terminal Cretaceous Hell on Earth, largely because weathering of minerals neutralized this acid onslaught. Earliest Tertiary life was different, without dinosaurs and many kinds of plants. Abundant fossil spores found immediately about the Cretaceous-Tertiary boundary beds are evidence that small weedy ferns flourished in the devastated landscape. These and other plant-like soil microbes built reserves of soil humus, robbing the air of carbon dioxide, and burying it in peat and clay. The ferns were followed by pioneering species of broadleaf trees and swamp cypresses, accelerating atmospheric restoration. Plant growth and mobilization of nutrients in soil solution also fertilized groundwaters and the sea, where phytoplankton blooms further reduced atmospheric carbon dioxide. Paleosols of the early Tertiary are not greatly different from paleosols of the late Cretaceous. Many fundamental components of ecosystems had survived. There were still many kinds of plants and animals. Although many of the old players were gone, including the once-dominant dinosaurs, the theater and much of its fundamental organization survived to build again on the ashes.

This is not the only example of a post-apocalyptic greenhouse warming in the history of the Earth. Studies of stomatal index of *Ginkgo* and related plants indicate some 13 such times of elevated carbon dioxide over the past 300 million years. These greenhouse crises vary in intensity, but each time carbon dioxide became more than about 2000 parts per million by volume, organisms died in large numbers, as at the Cretaceous-Tertiary boundary. The greenhouse crisis of the twenty first century has been a rise from 370 to 408 parts per million of carbon dioxide, so there is a long way to go to match devastations of the past. Such perturbations are the mortal enemy of the Proserpina Principle, because they catastrophically liberate carbon stored in plants and soils. Nevertheless, the natural ups and downs of Proserpinan cycles create ecosystems capable of rolling with the punches. The world has grown used to change, even abrupt change. Life is not rare or delicate.

DINOSAUR DEMISE

There was no warning, nor place to hide,

On the day of doom when dinosaurs died.

A fireball's flash on a distant shore

Preceded a crack and hot wind's roar.

The sudden sound and light on the face

Made many turn to that far-off place

To a column of dark expanding cloud

That stealthily wrapped all in a shroud

Of choking gas and stinging ashes

Settling quietly on dying masses.

There was no way to prepare for this,

And many fell to its deadly hiss.

7

An Occasion for Flowers

Flowering plants now have adaptations to insect pollination and animal dispersal, but those advantages were not useful to the first flowering plants. The earliest known flowering plants were in weakly developed paleosols where their abbreviated life cycle enabled them to colonize disturbed land.

Ancient Eleusis (now Elefsina), celebrated in classical Greece and Rome for its mysteries of Demeter and Persephone (Ceres and Proserpina of the Romans) is now an archeological preserve amid petrochemical plants of the outer suburbs of Athens. I visited in spring, to see the very place where legend has it that Persephone (Proserpina), the goddess of spring wildflowers, was abducted by Hades (Pluto). He emerged from, seized, and then dragged her down nearby caves, which overlook the coastal plain like eye sockets of a gigantic skull emerging from the earth. To my surprise, the twin caves still contained offerings of coins, fresh tangerines and unwilted flowers, indicating recent devotion. Ancient ways have survived 2000 years of eastern orthodox Christianity, represented by the Byzantine church on the top of the skull rock. The flowers of spring still emerge from cracks in the crags and ruins of this industrial quarter of Athens: large chrysanthemums (*Chrysanthemum coronaria*), pale wands of asphodel (*Asphodelus aestivus*), and nodding bell flower (*Campanula glomerata*). Flowers lift our spirits, but when they first evolved, they also transformed our world.

Flowering plants are everywhere now, on all kinds of soils, so it is difficult to imagine a world without them. When dinosaurs first roamed the Earth, flowers were scarce, perhaps non-existent. Although some fragmentary

© The Author(s), under exclusive license to Springer Nature
Switzerland AG 2022
G. J. Retallack, *Soil Grown Tall*,
https://doi.org/10.1007/978-3-030-88739-1_7

and enigmatic fossils as old as Triassic (230 million years old) could represent flowering plants, it is not until Cretaceous time (125 million years ago) that they appear convincingly preserved and diverse within the fossil record. Determining exactly when they arose is difficult because of the rarity of fossil flowers which show clearly the enclosed seeds from which comes their technical botanical name of angiosperm. Before angiosperms, Alfisol and Ultisol soils supported conifers and extinct seed plants with superficially fern-like leaves, rather than oaks and eucalypts. Before flowering plants, the principal invaders of young Entisol soils were ferns, such as royal fern (*Osmunda*), and horsetails, such as scouring rush (*Equisetum*), rather than grasses and daisies. Ferns, horsetails and conifers remain common in waterlogged soils such as Histosols, and low fertility soils such as Spodosols, but many flowering plants have adapted to these soils as well. Unlike grasslands that produced their own distinctive Mollisol soils, flowering plants modified soils in more subtle ways, for example, by creating leaves that rot to finer humus than masses of needles under conifer forests. Expansion of flowering plants from a few to almost all soil types is a remarkable case of evolutionary radiation. What was it about them or their environment that encouraged their rise to dominance in the green forests of the dinosaurs? Why were they so special? How did they change the world?

Could the success of angiosperms be due to the flowers themselves? The larger and livelier their petals, color, and fragrance, the better we like them. It takes a botanist's dedication and persistence to love the diversity of small-flowered plants, such as the grasses of East Africa or the many prickly shrubs of southeastern Australia. The flowers of gardens and florists today are the product of a worldwide search during the late nineteenth century for the most spectacular of the world's blossoms. Botanists enlivened the cold dank gardens of European colonial powers with exotic plants from the jungles of South America, central Africa, and southeast Asia, and from forests of the Pacific Northwestern United States, and foothills of the Himalaya. Stocks of rhododendrons from the Himalaya, roses from China, gladioli from Africa, and fuschia from South America were cultivated for the garden and vase. Artificial selection of cultivars for showiness and other features has produced flowers unparalleled in nature.

Human appreciation of flowers goes well back into prehistory. Roses were cultivated by ancient Greeks and Romans. Pollen extracted from a 60,000-year-old Neanderthal burial site at Shanidar Cave, Iraq, came from the most attractive wildflowers of the region, whereas pollen of grasses, oaks, and cedars common locally, were rare in these samples. Even at that remote time, early humans adorned graves with bouquets of fragrant and attractive flowers.

Flowers have an even longer history of coevolution with other animals. Flowers offer them food in the form of pollen and nectar, and enticement in their fragrance and distinctive shapes and colors. Benefit to the plant comes from the faithfulness of some animals for pollinating particular species of flowers. The dustings of pollen on the hairy legs of bees, the fur of bats or the feathers of birds are transferred to other plants, with minimal amounts wasted on other species. Animals offer plants a more efficient means of pollination than does broadcasting pollen on wind and water. The relationship between flowers and their pollinators is so close that one can talk of a syndrome of features that characterizes, for example, bee flowers. Bees frequent bilaterally symmetrical flowers like orchids and peas, which have a horizontal petal like a landing strip in front of a well of nectar. Beetles, on the other hand, pollinate bowl-shaped flowers that are radially symmetrical like magnolias. Flies are drawn to blossoms with pungent odor, like the arum lily. Birds with long beaks and butterflies with their coiled proboscis exploit deep tube flowers for nutritious nectar. Bats pollinate large night-blooming flowers of cacti and agaves. The close ecological ties between beasts and blossoms have been important for the evolutionary diversification of both. With faithful pollination comes restriction in the flow of genes, allowing the evolution of new species. Novelty, overwhelmed in a large crowd, makes a difference in small groups. The coevolutionary diversification of pollinators and flowers has produced the flowers and fruits that we value most, as well as a diverse array of birds and bees.

Fruits could also have played a role in angiosperm evolutionary success. Apples, oranges, peaches, watermelons, and grapes are all products of flowering plants or angiosperms. Feasts of mangoes by a Queensland track, persimmons in an Indiana woodland, grapes on a Greek hillside, and blackberries in a Scottish field all are gifts of flowering plants. The sweet pulpy flesh of angiosperm fruits feeds a variety of animals that take seeds beyond the oppressive shade of the parent plant. The smaller seeds are eaten with the pulp and are resistant to stomach acids. They are carried off in the guts of animals and deposited in a moist, warm, fertilized pile. Animals also disperse spiny and barbed fruits that stick to fur and skin. Others, such as the seeds of passionfruit, are sticky and adhere to feathers, feet, and faces. As for pollination, dispersal of fruits also involves cooperation between animals and plants. Is this why there are so many species of flowering plant in the world today?

Such a bounty of fruits and flowers so entices our senses that they tend to dominate our thinking about flowering plants, but grasses and plantain, sheep sorrel and daisies, are also flowering plants. With them we are at war with mowers and herbicides, for these are weeds. Grasses and daisies are the worst of weeds because of some seminal features of angiosperm reproduction. Conifers, ferns, and horsetails also include species that are weedy in the sense that they regrow after damage, quickly colonize vacant ground, and produce millions of small, widely dispersed propagules. Unlike conifers, ferns, and horsetails, however, flowering plants have no delay between pollination and fertilization. Reproduction is fast, a matter of minutes. Seeds can mature, be dispersed, and germinate within a matter of days in weedy flowering plants. Weeds germinate whenever conditions allow. A little opening in the grass, some moisture and sun, and a whole new crop of weeds grows through the lawn. Their tiny seeds waft in on parachutes of down, on bird's feet, and on the wind itself. Their pollen spreads in yellow clouds. The blanket of tiny propagules covers the land, waiting for opportunity. Reproduction is mainly what they do. No energy is wasted on producing fleshy or stony fruits, on petals or nectar, on large leaves or tree trunks. They may not be as pretty as the roses, azaleas, and rhododendrons that we are trying to protect from weedy growth, but flowering weeds deserve admiration for effort. Did such weedy plants play a role in the early evolution of flowering plants?

Before flowering plants, much vegetation was like my current home amid the forests of the Pacific Northwestern U.S., dominated by conifers, horsetails, and ferns. Oregon forests endure two long seasons, a dry summer and a rainy winter. Spring blossoms of rhododendron and thimbleberry lift the spirits after months of living within gloomy green firs. Did the first of Cretaceous flowers that greeted the dinosaurs present such a show of blossoms? It is an appealing image, appearing occasionally in the rapidly burgeoning genre of dinosaur art: a duckbill dinosaur munching vacantly on a bush adorned with showy *Magnolia* flowers. There were indeed some showy flowers associated with the dinosaurs, but the fossil record of flowering plants and soils indicates a more humble origin of flowering plants. Flowering plants crept in like weeds of the lawn, rather than the cosseted ornamental blossoms of the garden.

Evidence of the earliest flowering plants has been slow to accumulate because flowers are uncommon as fossils. Flowers bloom for only a few days to weeks of the year. Chances are slim that they will fall into a river or lake, and be preserved intact among the much more numerous leaves and twigs of plants. Finding a needle in a haystack is a good analogy for finding fossil flowers. In recent years, however, the available haystacks have been sifted for fossil flowers by paleobotanists, such as David Dilcher of the Florida Natural History Museum, Else-Marie Friis of the Swedish Natural History Museum, and Peter Crane of Kew Herbarium, England. As a result the fossil record of flowers is becoming increasingly better understood. The earliest fossil flowers

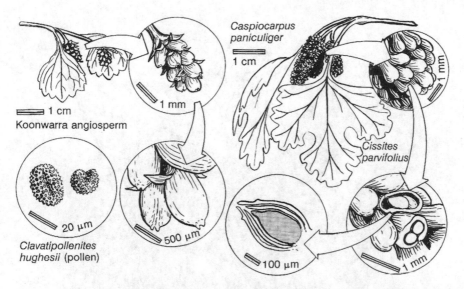

Fig. 7.1 Reconstructions of Early Cretaceous flowering plants including an unnamed fossil (left), from Aptian (117-million-year-old) rocks of Victoria, Australia, and a middle Albian (105-million–year-old) plant from Kazakhstan. These very ancient flowers of small weedy shrubs have abundant small pollen and seeds like modern weeds, rather than the showy petals or nectaries of modern garden ornamentals

of the Cretaceous were not magnolias or roses, but extinct plants that had neither petals nor nectaries, nor succulent stone fruits. Instead, they were small flowers and produced numerous, small seeds, and copious, small, pollen grains, like modern weeds (Fig. 7.1). By later Cretaceous time, there was a greater variety of flowers and fruits, including some that we would have found attractive. Nevertheless, there were still petal-less flowers arranged into balls and tassels (Fig. 7.2). Weeds rather than ornamentals appear fundamental to the early evolution of angiosperms.

The paleosols in which early flowering plants were found tell a similar story of weedy origins. Early Cretaceous paleosols crop out in road cuts of the rolling hills of Victoria in Australia, the mountains of British Columbia in Canada, and the humid piedmont of Maryland, U.S.A. The earliest flowering plants in each of these regions are found in thin paleosols, with bedding planes mixed little by fossil roots. These Entisol paleosols are like soils of streamsides and lake or ocean margins. Such areas are frequently disturbed by floods and storms. Other early angiosperm fossils from lake beds of Victoria (Australia) and Liaoning (China) do not have associated paleosols, but may have been from lake-margin Entisols. Conifers, ferns and cycad-like plants dominated coal-bearing Histosol paleosols of ancient swamps and some of the gray Inceptisol paleosols of seasonally inundated lowlands. Unlike these

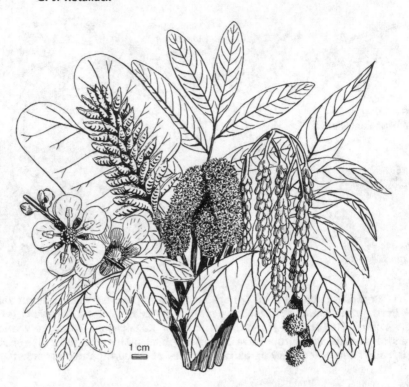

Fig. 7.2 A mid-Cretaceous (Cenomanian, or 94-million-year-old) bouquet of extinct flowering plants from Kansas, U.S.A.: clockwise from left, leaves of *Pabiania variloba* and associated *Dakotanthus cordiformis*, leaf of *Liriophyllum kansense* and its likely fruit *Archaeanthus linnenbergeri*, leaf of *Sapindopsis bagleyi* and its pollen-bearing catkins, leaves of *Pandemophyllum kvaceki* and its likely fruit *Prisca reynoldsii*, and *Eoplatanus serrata* leaves and their fruiting heads *Friisicarpus dakotensis*. Paleobotany has a tradition of naming fossil flowers separately from leaves because of uncertainty about which belong together, and because of the different evolutionary rates of different plant organs

drab paleosols with their associated fossil plants, the formerly well-drained Alfisol and Ultisol paleosols of the Early Cretaceous are massive red claystones with deeply reaching, poorly preserved, root traces, but devoid of identifiable plant fossils. It is unlikely that flowering plants were common in this upland vegetation, because throughout the Early Cretaceous, ferns and conifers are much more abundant than angiosperms in the regional pollen-rain in lake and marine shales. Flowering plants were locally common but regionally rare, as is generally true of weedy plants.

Even by mid-Cretaceous time some 91 million years ago, when a variety of flowering plants had evolved, their predilection for disturbed sedimentary environments is still clear from badlands and brickpits in the Dakota Formation of Kansas and Nebraska. As a postdoctoral assistant of David

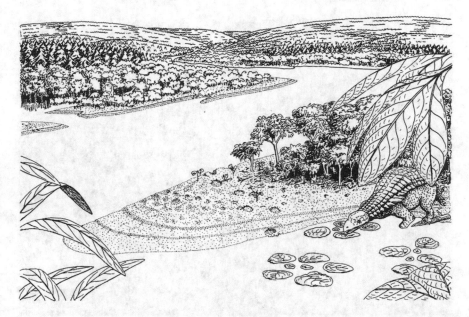

Fig. 7.3 Reconstruction of a mid Cretaceous (94 million years) streamside in what is now central Kansas, U.S.A. Most of the hills and floodplain are overgrown with conifers similar to coastal redwoods (*Sequoia condita*), but the river margins support mostly angiosperms including water lilies (*Aquatifolia fluitans*) and woody shrubs of a sycamore-like plant (*Araliopsoides cretacea*). The dinosaur is an early ankylosaur (*Silvisaurus condrayi*)

Dilcher, then at Indiana University, I made several pilgrimages in search of early flowering plants and the soils in which they grew. In mid-Cretaceous time, central Kansas was a humid subtropical coast to an enormous mid-continental seaway that stretched from the Gulf of Mexico to the Canadian Arctic. The sea shells and chalk of that ancient seaway still litter the prairie. There are extinct ammonites (*Acanthoceras*), coiled like ram's horns, large mussels (*Inoceramus*), ridged like the ripples on a pond, and familiar kinds of gooseneck barnacles, clams, and snails. On the margins of this shallow sea were deposits of coastal lagoons, lakes, and rivers, as well as a variety of ancient soils, which can be used to reconstruct regional patterns of vegetation (Fig. 7.3).

Fossilized soils formed closest to the sea are black with organic matter and bright with pyrite, like Histosol soils of modern intertidal woody vegetation, or mangal. Such an interpretation is supported by the discovery of mussels (*Brachidontes*) with both valves articulated and erect, preserved in their original life position within the paleosols. Comparable bivalves are found around the roots of mangroves and salt marsh grasses of the southeastern U.S. today. The fossil mangrove plants of these mid-Cretaceous paleosols were

Color Photo 7.1 Mid-Cretaceous (Cenomanian, 91 million years) Cahli paleosol (Histosol) with tree trunk cast, and thin tuff with angiospem leaf fossils in Dakota Formation near Westwater, Utah

primarily flowering plants with thick leathery leaves (*Pabiania*), but were not closely related to living species of mangroves. The extinct mangrove flowers (*Dakotanthus*) had attractive blossoms with five petals and anthers, and produced only about 10 moderately sized seeds.

Also dominated by flowering plants are fossil leaf litters in Histosol paleosols, now represented by coal seams, formed in freshwater swamps, where pyrite and clams are rare. Swampland trees are evident from tree trunk casts in the coal (Color Photo 7.1). These angiosperms had larger and thinner leaves (*Liriophyllum*) than the fossil mangroves. The flowers of these swampland flowering plants included large petals and helically arranged follicles similar in some ways to those of living *Magnolia*.

Other flowering plants again, comparable to living sycamore (*Platanus*), were found in gray shales and thin clayey Entisol paleosols of lake and lagoon margins. Other lobed leaves (*Araliopsoides*) dominate sandy Entisol paleosols of ancient streamsides (Color Photo 7.2). The fossil fruits of these Entisol angiosperms are balls and tassels producing abundant small seeds and pollen grains, more like weedy plants than the fossil mangrove and swamp flowers.

No fossil plants were found in the red, deeply weathered Ultisol and Oxisol paleosols of upland forests. These well-drained forests were probably dominated by conifers, considering that the regional pollen flora entombed in the

Color Photo 7.2 Mid-Cretaceous (Cenomanian, 91 million years) Sagi paleosols (sandy Entisols) with angiosperm leaf fossils in the Dakota Formation near Hoisington, Kansas

Dakota Formation contains mostly conifer pollen. By mid-Cretaceous time, flowering plants had made great inroads into disturbed sedimentary environments, and included both weedy and ornamental plants, but had not yet displaced conifers in stable parts of the landscape.

Modern deeply weathered soils comparable to those of the ancient conifers in the Dakota Formation now support tropical rain forest consisting almost exclusively of flowering plants, and coevolved with a variety of insect, bird and mammal pollinators. Such tropical forests of flowering plants evolved much later, some 45 million years ago. I know them best from fossil soils and plants of the Eocene age, Clarno Formation of central Oregon, where I take university classes on annual field trips. The paleosols found there are thick, clayey, red, and deeply weathered compared with their volcanic parent materials, and are like Ultisol and Oxisol soils on the flanks of subtropical volcanoes of central America. In the Clarno Nut Beds, so called for their fossil fruits and seeds, there are common remains comparable to the ornamental flowering plants of a florist shop: large palm seeds, shells and meats of walnuts, small *Magnolia* seeds, and pitted stones from the succulent fruits of tropical vines. The diversity of plant remains in these Eocene deposits is also impressive: 173 species of fossil plants is not remarkable for a modern rain forest plot, but is exceptional for a single fossil plant locality. Many of these fossil plants are related to flowers pollinated by bees today. Although fossil bees have been found as ancient as late Cretaceous, the evolutionary radiation of bees and bee flowers is primarily an Eocene phenomenon. The coevolution of angiosperms with pollinators and dispersers was a longer term and later process than the original early Cretaceous rise to dominance of weedy flowering plants.

If the Cretaceous floral revolution was a time of wide dispersal of flowering weeds, rather than ornamentals, what were the causes and consequences of this revolution? The most ancient pollen of flowering plants is found in 120-million-year-old deposits of Gabon and Brazil. During that part of the Cretaceous period, coastal Gabon and Brazil formed a rift valley that preceded the opening of the South Atlantic Ocean. From this tropical homeland, early flowering plants spread north and south to both poles and to every continent over a period of about 20 million years during the Early Cretaceous. The global spread of these early weedy plants may have been aided by dramatic oscillation in sea level characteristic of Cretaceous times. Vast continental seaways spread not only across North America, but across the current Sahara Desert, well into Brazil and to the very interior of Australia. This high sea level was at a time of warm global climate, as indicated by fossil forest stumps are preserved at very high paleolatitudes. Such profound geographic reorganization with the coming and going of shallow seas left large tracts of low-lying land open for colonization. Flowering plants with their abbreviated reproductive cycle could flower and set seed more quickly than other plants. Sea shores and rivers provided highways for global spread of tropical weeds in a generally warming global climate that culminated in the mid-Cretaceous.

Trampling by dinosaurs may have been another factor in the success of flowering plants. Elephants today give East African game parks a decidedly vandalized appearance of broken trees, trampled bushes, and muddy trails. Large dinosaurs also would have been rough on vegetation. In the Early Cretaceous when flowering plants were becoming diverse, there were new kinds of dinosaurs with long batteries of teeth, cheek pouches, and low browsing range capable of processing large quantities of low vegetation. Remains of these duckbill, ankylosaur, and ceratopsian dinosaurs are prominent in association with fossil flowering plants and their former soils. In some regions these newer kinds of dinosaurs supplanted the archaic long-neck dinosaurs, or sauropods. Evidence from fossil trackways in the soils they trod indicates ecological separation of ancient and new dinosaur groups. Duckbill and ceratopsian dinosaurs are associated with gray Entisol and Inceptisol paleosols of the flowering plants. Sauropod trackways, in contrast, are found associated with red Aridisol and Alfisol paleosols which probably supported conifers. This separation of conifer-sauropod and angiosperm-duckbill communities persisted throughout the Late Cretaceous. Flowering plants coped better than other plants with the large herds of low-browsing duckbills, ceratopsians, and ankylosaurs. Global warming from heavy browsing pressure of dinosaurs, according to the Proserpina Principle, may at first have promoted global spread of tropical flowering plants. Warmth peaked during the mid-Cretaceous when palms and tropical soils were as far north as Kansas. This Cenomanian warm spike of 91 million years ago, was one of a series of global warm spikes like that at the Cretaceous-Tertiary boundary of 66 million years ago. These warm spikes of high carbon dioxide may also have had similar catastrophic causes from impacts or large volcanic eruptions. These balmy conditions were interspersed with cooler times including evidence of Early Cretaceous sea ice, and frigid marine crystals of glendonite in Australia.

Rapid regeneration and decay to humus of flowering plants after floods and dinosaur trampling later in the Cretaceous may have had chilling consequences, again according to the Proserpina Principle. Many paleosols of Cretaceous flowering plants are more carbon-rich than those of conifers. This remains true if comparison is limited to paleosols with comparable depth of rooting and of calcium-rich nodules, indicating former good drainage and comparable rainfall regimes. Alfisol paleosols of conifers and sauropods in the Late Jurassic Morrison Formation of Utah have nodules at a depth of 3 ft. (1 m) and deeply penetrating root traces, but are red with iron oxides and very poor in preserved organic matter. Inceptisol paleosols of flowering plants and duckbill dinosaurs in the Late Cretaceous, Two Medicine Formation of Montana also have deeply penetrating root traces and nodules at a depth of a bit more than 2 ft. (50 cm), indicating well-drained soils of an even drier paleoclimate with less moisture to leach calcium deeper into the profile. Despite this, paleosols associated with fossils of duckbill dinosaurs and flowering plants are gray with organic matter, in contrast with the archaic red soils of sauropods and conifers. Comparable differences persist today. Pine plantations with their resinous leaf litter, and red, sandy soils, can be contrasted with nearby oak forest growing in brown-gray soils with well humified leaf litter and clayey soils. On a global scale, improved carbon storage in angiosperm-dominated soils, and in shales eroded from soils, had chilling effects.

These observations suggest a new view of the consequences of flowering plant expansion through a variety of soils. By colonizing disturbed ground rapidly with leafy green carpets, and by enriching the soil with humus, flowering plants could have begun to draw down the great carbon dioxide greenhouse of the Cretaceous. Carbon dioxide from the atmosphere is not only consumed as plant matter and organic soil, but by forming carbonic acid in soil water. This weak acid attacks mineral grains, releasing nutrient elements to plants, while exporting carbon in bicarbonate of groundwater. Organic soils, formed under easily decaying leaves of flowering plants, yield carbon-rich sediment for burial in lakes and oceans as a long-term disposal of carbon that would otherwise thicken the air with carbon dioxide. The quicker and denser colonization of new ground by angiosperms, compared with slower regrowth of other kinds of plants, would have accelerated all these mechanisms of carbon burial, with potential global consequences. The earliest tropical flowering plants profited in their global spread from generally warm climates and fluctuating shorelines of the Early Cretaceous, but as flowering plants became widespread in the Late Cretaceous, alpine glaciers expanded and continental seas retreated.

Flowering plants were unwitting, but nevertheless effective engineers of global change. Angiosperms modified existing soils, rather than creating new soils like grassland Mollisols, but even these modifications may have had environmental consequences. In the 1960's many young people like me embraced the hippie movement, as a return to kindness, peace, nature, and living in the moment. The aphorism of the decade was "flower power". Flowers do indeed have the power to change both our inner world and our outer environment.

FLOWER POWER

Spring rhododendrons burst into bloom

Amid Oregon's dank firs none too soon.

Pink and white blossoms burden their shoots,

Lure crowds of bees, hint at summer fruits,

But grasses and plantains flushed with green

Have flowers too small to be easily seen.

Everywhere now are plants with flowers,

But dinosaurs first trod bloomless bowers.

The earliest flowers grew on small weeds,

Dusty with pollen and quick to set seeds.

A world conquered by weeds later adorns

With magnolias, orchids, figs and acorns.

8

Dinosaurs and Dirt

Many dinosaur fossils are found in paleosols which are clues to their life style and evolutionary origins. Dinosaurs began as small predators in forest paleosols. Long-neck dinosaurs were browsers in paleosols of dry woodlands.

Most children go through a dinosaur phase. They read avidly about them, draw or color them, or place plastic dinosaur models in sculpted landscapes in the sand-box, or garden. Some of us never grow out of our fascination with dinosaurs and their world, but there is more to dinosaurs than the icon of a plastic figurine, or trading card. Dinosaurs were once living creatures and parts of ancient communities. As with the dinosaurs themselves, their world remains strange and elusive. Were dinosaurs like oversized crocodiles or big birds? Were their plant communities like African grasslands or Asian deserts? The paleontologist Everett Olson called these kinds of questions "me too paleoecology." By this he meant the assumption that ancient organisms and their communities were similar in most ways to those of the present. It is as if ancient ecosystems can be compared with a checklist of modern ecosystems. None of the modern analogs for dinosaurs have worked especially well because each dinosaur was yet another unique species in an extinct community. The problem became most apparent to me when I was approached by staff of the BBC-television special *"Walking with dinosaurs"* to recommend suitable habitats as a background to their computer-generated dinosaurs. We drove for many miles in Oregon and found very few sites untainted by modern trees, grasses, or flowers. Nor could we find modern soils comparable to soils of dinosaurs, which were probably more exposed to view than modern

G. J. Retallack, *Soil Grown Tall*, https://doi.org/10.1007/978-3-030-88739-1_8

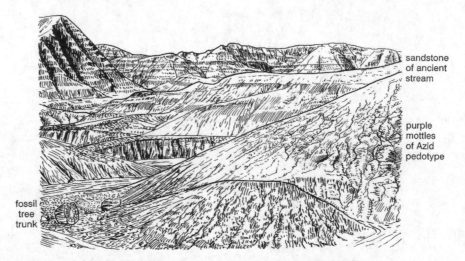

Sandstone labels on figure: sandstone of ancient stream; purple mottles of Azid pedotype; fossil tree trunk

Fig. 8.1 Purple-mottled horizons of these badlands are formerly forested Alfisol paleosols of Late Triassic age (230 million years old) in the Blue Mesa area of Petrified Forest National Park, Arizona, U.S.A.

soils because of dinosaur trampling and because ground cover of aridland lichens, mosses and ferns would have been less effective than later-evolved grasses. Evidence from fossil soils allows a vantage point independent from the fossils, from which we can reassess the origin and rise to dominance of dinosaurs, and their role in another grand swing of the Proserpina Principle.

The most ancient known dinosaurs are of Late Triassic age (230 million years old). Among these, the best known are turkey-sized *Coelophysis*. In 1947, Ned Colbert discovered a herd of complete skeletons near Ghost Ranch, New Mexico, in red beds of the Chinle Group. These spectacular fossils can still be seen in the American Museum of Natural History in New York. These fossils represent well the early dinosaurs: small to medium sized, slender, swan-necked, agile predators. The paleosols in which they have been found reveal much about their past lives.

Arizona's Chinle red rock country is the quintessential landscape of American western movies, and the paleosols that make it red are thick (1 m or 3–4 ft.), clayey, and sometimes calcium-rich. The red, clayey main portion of each paleosol is capped by purple and dark gray mottled upper parts, an overall profile form like that of a modern Alfisol soil (Fig. 8.1, Color Photo 8.1). These mixed colors create rainbow rocks of unusual scenic charm, well exposed in scenic badlands of Petrified Forest National Park, Arizona. The huge agatized tree trunks preserved here have been thought to be log jams within ancient stream deposits, but I am not the only one who has excavated

Color Photo 8.1 Late Triassic (Norian, 222 million years) Azid paleosols (Alfisols), in Blue Mesa Member, Petrified Forest Formation, Petrified Forest National Park, Arizona

fossil stumps and their downward penetrating roots to come to the conclusion that many of these agatized wood assemblages are true fossil forests. The thick, red, clayey paleosols are littered with large logs and riddled with petrified roots and stumps. Modern Alfisols too are typical of forests. Most of the wood is red and petrified, literally "made into stone", but some is brown to black with original cell walls permineralized with silica. Under the microscope, the permineralized wood of the fossil trees is similar to that of living Norfolk Island pine of Australia or Arauco pine of Chile (both genus *Araucaria*). Associated fossil cones are unlike those of these living trees, so the Triassic trees were extinct species. A paleoclimate drier than usual for Arauco and Norfolk Island pines today, can be inferred from the kinds of clay and depth to calcite nodules in these fossil Alfisols, which are evidence of a subhumid climate of 32–39 in. (800–1000 mm) mean annual rainfall. The stumps of the fossil trees also were more widely spaced than in modern Arauco forests, another indication of subhumid climate. There also are distinctive chambered nests in these paleosols, similar to those of modern ground-dwelling termites of the seasonally dry tropics. The termite nests indicate a warm subtropical to tropical climate for Arizona during Triassic time.

These open Triassic forests had a combination of plants, animals, and climate unlike any today.

The earliest dinosaurs, *Coelophysis*, may have lived in these Triassic forests, but were not preserved in forested paleosols. At Ghost Ranch in New Mexico, their articulated skeletons litter Entisol paleosols that have retained bedding planes, as evidence of weak development and a short period of soil formation. The herd of dinosaurs probably died in a small stream wash within the forest, perhaps during a flood. It requires rapid burial to preserve fully articulated skeletons from the scatter and dissolution that is the fate of most carcasses in the wild.

Paleosols reveal other kinds of soils and ecosystems on these Triassic landscapes of Arizona, in addition to reddish-brown, forested Alfisols and streamside, young Entisols (Fig. 8.2). Purple-gray, mottled, Alfisol paleosols have the subdued gray colors and spreading roots of soils under seasonally waterlogged lowland forests, and contain bones of extinct crocodile-like creatures called phytosaurs. Gray Inceptisol paleosols with decayed plant remains and calcite nodules have yielded a few loose teeth of another small reptile (*Tecovasaurus*) and many bones of the last of the mammal-like reptiles: odd-looking creatures which have been called *Placerias*. They were herbivorous, squat animals with tusks and beak. Dinosaur expert Dale Russell calls them cow-turtles, a description apt for both their appearance and habitat inferred from paleosols. The abundance of their remains in gray paleosols indicates a preference for swamps and wallows. Other gray Entisol paleosols, with much evidence of youth, such as bedding and ripple marks barely disrupted by roots, have yielded fossils of extinct tree ferns and large horsetails. These represent waterlogged swamp and river margin soils, but have not yet yielded remains of fossil reptiles. A variety of paleosols reveal an array of floodplain habitats and communities, but few of these communities included dinosaurs, which remained rare.

Judging from the distribution of bones in paleosols, the early dinosaurs, *Coelophysis* of New Mexico and Arizona, were active, agile predators of forests and streamsides. Their lightly skeletonized skulls full of sharp teeth, their erect bipedal stance and modest size, suited them well to forests full of insects, and small ancestors of mammals. The large herbivores of their time were mammal-like reptiles such as *Placerias* and other archaic reptiles. The earliest dinosaurs were not the top predators of their time. That role was filled by extinct reptiles called erythrosuchids, about the size and shape of the living Komodo dragon of Indonesia. Early dinosaurs were not aquatic predators of fish and crustaceans. Crocodile-like phytosaurs and a variety of other reptiles

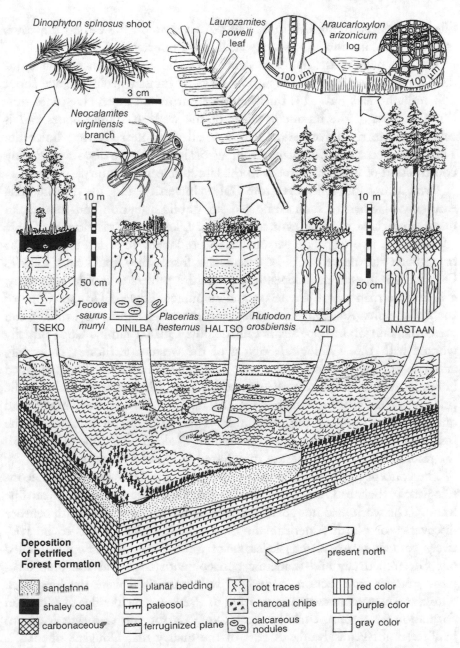

Fig. 8.2 Reconstruction of soils, plants and animals of Petrified Forest National Park, Arizona, during Late Triassic geological time, some 230 million years ago. Both the small reptile *Tecovasaurus* and cow-turtle *Placerias* have been found in Dinilba Inceptisol paleosols, but the crocodile-like phytosaur *Rutiodon* in Azid Alfisol paleosols. The other paleosols included Histosols (Tseko), Entisols (Haltso) and Alfisols (Nastaan) (reprinted from Proceedings of the Dinofest Conference, 1997, with permission of Dinofest International and Drexel Univerity, Philadelphia)

filled that role. Among this throng of extinct reptiles, the earliest dinosaurs were neither common, diverse, powerful, nor important.

The true reign of the dinosaurs came later, and is well known from abundant bones of another set of rainbow rocks of Jurassic geological age about 153 million years old. This thick band of purple-red-green claystones runs through most of the mountain states of the western United States and is called the Morrison Formation after a small town near Denver, Colorado. The most spectacular finds came from Wyoming and were discovered during construction of western railroads in the late nineteenth century.

Rival teams of collectors from Yale University and the Philadelphia Academy of Sciences soon competed for public attention and funding in unearthing these startling extinct creatures. Like Aesop's classic tale of hare and tortoise, two disparate personalities were locked in a race to be first to bring back and name the most spectacular fossils they could find. Charles Othniel Marsh was an independently wealthy professor at Yale University, a deliberate man of great tenacity and resources. Edward Drinker Cope of the Philadelphia Academy of Sciences, on the other hand, was brilliant and mercurial, son of an upper-class Philadelphia Quaker family. Cope initially worked with U.S. Geological Surveys of the western territories, collecting, and establishing contacts as far west as Oregon. Hired collectors shipped bones in railroad cars to both men. Each man tried to establish scientific priority by erecting new names for the fossils. Mistakes were made, and quickly seized upon by the other man. Their public feud as much as the fossils themselves fueled public enthusiasm for paleontology, unequalled until the recent renaissance of interest manifested in "Jurassic Park" movies.

The concept of dinosaurs was invented as late as 1842 by British paleontologist Sir Richard Owen from two ancient Greek words meaning "terrible lizard." The world had not yet seen anything like the late nineteenth century discoveries of North American dinosaur fossils, which proved to be relatively complete, articulated skeletons of astonishing diversity. They filled out Owen's sketchy and academic concept with large and fearsome creatures stranger than fiction. Especially well publicized were the long-neck dinosaur skeletons excavated from 1909 to 1924 on a rocky ridge of Morrison Formation in eastern Utah by Earl Douglass for the Carnegie Museum in Pittsburgh, Pennsylvania. A part of the quarry that Douglass opened is still preserved for visitors in Dinosaur National Park, astride the border of Utah and Colorado. The long-neck dinosaurs are commonly referred to as "*Brontosaurus*," but the fossil skeleton used as the basis of that name by Marsh was later shown to belong to the same species as fossil vertebrae, limbs, and tail bones of another individual, which he had named *Apatosaurus* only two years earlier in 1877. Recently the name *Brontosaurus* has been revived for a rare form of dinosaur including the type specimen of Marsh, but *Apatosaurus* is the more common genus of these shovel-nosed long-necks.

Four main kinds of these incredible long-necked dinosaurs, or sauropods, have been found in the Morrison Formation. In order of abundance, they are *Camarosaurus*, *Diplodocus*, *Apatosaurus*, and *Barosaurus*. Each is distinct in the relative elongation of its skull, its teeth and other features. What can be made of these strange chimaeras, these beasts with bodies like elephants, tails like lizards, and necks like giraffes? They had the blunt teeth of herbivores, yet a head that seems too small to feed such a large body. Their nostrils were high on the skull, like a crocodile's. How did they manage?

One early idea was that sauropods lived in lakes. By this view, the high nostrils were considered an adaptation to life in water, which would also buoy up their tremendous bulk. Unfortunately, the water pressure at depths equivalent to their long necks would have made it impossible for them to breathe underwater. Engineers have taken another look at the backbone of these creatures, which is composed of enormous vertebrae with hollow spaces between trusses, like those of a steel bridge. It was a well designed support system. Fossil footprints have been found as evidence that sauropods walked on dry land around lakes. The skeletons found by Earl Douglass were in an ancient river channel, where carcasses became stranded after floods. Some of these carcasses were torn apart and chewed by carnivores, such as the large predatory *Allosaurus*. Sandstones remaining from the stream channels have the shallow spreading geometry and cracked surfaces of intermittent and seasonally dry streams, implying that water was not always there for them.

Evidence that sauropods lived on dry soils, rather than lakes, comes from associated red, clayey paleosols. Late Jurassic paleosols of the Morrison Formation were Alfisols and Aridisols, studded with fine-grained, calcite nodules, which preserved bone well (Fig. 8.3). Commonly the nodules are some 2 ft. (50 cm) below the surface of the paleosol. Such a shallow accumulation of nodules is found now in soils of subhumid climates, receiving only 25–35 in. (600–900 mm) rainfall per year. Some of the nodules are elongated around large woody root traces. Other woody root traces have wide grey-green haloes. The stout roots and nodular, little-weathered paleosols are indications of former dry woodland. Woody vegetation is confirmed by rare permineralized seeds and shoots of conifers and cycad-like plants found in the paleosols. Large fossil termite mounds also have been found in paleosols of the Morrison Formation. The small, pale insects may have kept the soil clean of leaves and twigs, as they do in termite-infested soils of West Africa and northern Australia today. The Jurassic dry woodland probably had a "lived in

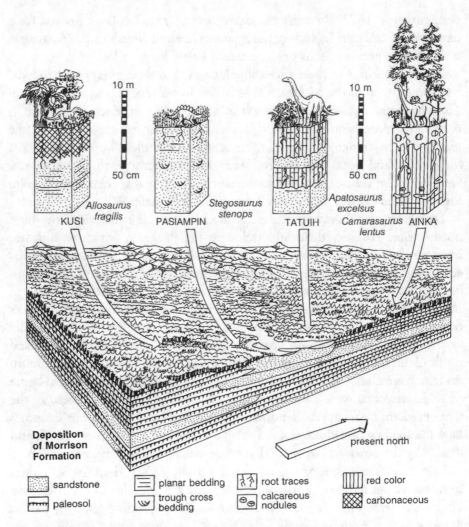

Fig. 8.3 Reconstruction of soils, plants and dinosaurs of the Morrison Formation at Dinosaur National Park, Utah, during late Jurassic time, some 153 million years ago. A variety of habitats is indicated by the different paleosols, but the dinosaur fauna of each is not as restricted as in Triassic paleosols. Jurassic dinosaurs were large creatures with large home ranges (reprinted from Proceedings of the Dinofest Conference, 1997, with permission of Dinofest International and Drexel University, Philadelphia)

look" of dismembered and broken trees, with open dusty patches, like African woodlands after the passage of a herd of elephants. However, there were no grasses in the Jurassic, and paleosols of the Morrison Formation lack the dark brown color, small crumb clods and abundant slender roots characteristic of grassland soils. This is a big difference from grassy East African game parks, which are often compared with dinosaur environments. A better modern analog for the environment of sauropods dinosaurs is the dusty, dry, native cypress-pine (*Callitris*) woodland of outback New South Wales, Australia, in which shapely conical trees of modest height are scattered in open formation, with intervening bushes, and the most prominent animals are wallabies and kangaroos. A herd of long-necked giants shuffling through dusty glades would have been an arresting sight, unlike any available today.

Sauropod fossils in such dry open vegetation pose other puzzles. How did they get enough to eat? They probably were giraffe-like browsers of dry woodlands, and like elephants would have had to eat almost constantly to maintain their great bulk. Unlike elephants, with their broad mouth, large flat-topped teeth, and trunk, sauropod dinosaurs had small mouths and peglike teeth. These pegs would have been little use in breaking down food. Grinding of their food may have been undertaken in a gut filled with stones. Several sauropod skeletons have been found with a cluster of polished pebbles where the stomach would have been. Like parakeets, they evidently swallowed stones that acted to break down food as a kind of gastric mill. Rock shops all over the western U.S. sell polished stones as dinosaur stomach stones. Beware that very few, if any, of these have been collected from an actual dinosaur skeleton!

Such voracious appetites would have locally outstripped local plant production, especially if they roamed in large herds, as is indicated by locally abundant fossil footprints of sauropods in the Morrison Formation. This problem would be lessened if sauropod dinosaurs had a lower basal metabolic rate than for elephants and thus a lesser need for food. Elephants, like all mammals, are warm-blooded and so expend much of the energy from their food in maintaining body temperature. There has been much debate in recent years about whether dinosaurs were warm-blooded like mammals, cold-blooded like reptiles, or something between the two. The great sauropod dinosaurs have spongy bone more like that of mammals than of crocodiles. This spongy bone, readily inspected in bins of dinosaur bone in many rock shops and nature stores, is from a complex system of bone-marrow cells. Such strongly vascularized bone would have allowed rapid growth rates to near-adult size from an embryo within eggs no bigger than footballs. Another line of evidence comes from oxygen-isotopic compositions which vary with temperature during bone growth. The oxygen isotopic compositions of bones from the chest compared with the tail bones of sauropods are uniform, as in the skeletons of warm-blooded elephants, not varied as in cold-blooded crocodile skeletons. If we can assume from the quality of bone-preservation that oxygen-isotopic composition of the bone was not altered during burial, this is evidence that temperature varied little throughout the bodies of dinosaurs. However, such a large creature in a near-tropical climate would not have varied much in temperature, even if as dependent for heat from the environment as crocodiles. This is because temperature in tropical regions varies little through the year, and the temperature difference between night and day is not sufficiently prolonged to outpace the thermal inertia of a large body. Sauropods also lack the nasal turbinals (scrolls of bone within the nasal cavity), and the large nasal cavity of many warm-blooded mammals, which adjust inhaled air temperature to lung temperature. This line of evidence is also insecure because some warm-blooded animals, including humans, have very small nasal turbinals. However we look at it, long-neck dinosaurs had a unique metabolism for such large creatures, not exactly like cold-blooded crocodiles, nor exactly like warm-blooded mammals either.

Even if we assume a fuel-efficient, cold-blooded metabolism for sauropods, there is still some doubt that dry woodlands would have produced sufficient forage for the herds. Herds of elephants on what is picturesquely called "the skeleton coast" of Namibia, survive by long migrations to exploit seasonally available riparian vegetation of this desert region. Such migrations are plausible for sauropods. As in Namibia, passage of the herd was probably hard on vegetation, diminishing its role in the carbon cycle.

This real "Jurassic Park" was thus unique in many ways, especially in its severe plant destruction and soil trampling. Colorado, Utah and Wyoming are now seasonally snowy, with desert rangelands and thin dry soils. In contrast, Jurassic paleosols are thick and clayey, and indicate not only subhumid, but subtropical climate. Fossil plants and paleosols have also been found near both the north and south poles of the Jurassic period. High-latitude plants and soils of the Jurassic were unlike modern periglacial and permafrost ecosystems and soils, and more like those of temperate climatic regions. Polar regions during the Jurassic may have been seasonally snowy, because fossil wood shows strong, asymmetric growth rings, and paleosols are thinner and less weathered than in paleotropical regions. No glacial tills, ice-cracks, or tundra plants have been found to indicate permafrost or polar ice caps. During the Jurassic greenhouse spikes, temperate paleoclimates extended into polar regions that are now barren ice caps.

The Jurassic greenhouse may also be reflected in paleosols of the Morrison Formation, because its paleosols have little organic matter, only small patches of gray color and coarse soil structure compared with soils now forming in similar warm subhumid regions. Jurassic paleosols are distinct in this respect also from the grey-purple paleosols of subhumid terrains that formed earlier during the Late Triassic, and later during the Late Cretaceous. Carbon storage in Jurassic soils and sediments was significantly less than for preceding and following periods of geological time. Carbon was liberated as carbon dioxide in the air by rapacious herds of dinosaurs, as well as by termites and other soil animals. The pattern of global atmospheric change here is the familiar Proserpina Principle, but on an evolutionary time scale of millions of years, rather than the annual alternation of photosynthesis and respiration. When plants were winning a greater share of the world's carbon by means of refractory or toxic materials, such as lignin of tree trunks and roots, or the living carpets of grassland, glaciers expanded into ice caps in a well oxygenated atmosphere. Later when animals such as large dinosaurs and termites with their unique gut microflora found ways to release that carbon, a balmy greenhouse prevailed. The great machine extending these biological feats of atmospheric and climatic change was probably the soil itself.

GREENHOUSE ERA

Mesozoic climate was unusually warm

With carbon dioxide above the norm.

Forests grew much nearer the poles

Where now soil's frozen with icy holes.

Gone were ice caps, floating glaciers,

Calving icebergs, marine till layers.

Carbon in trees was no longer secure

From rapacious feeding dinosaurs.

Carbon of litter fueled great masses

Of termites who made it soil gases.

During those times of humus-poor land,

Animals, not plants, had an upper hand.

9

World's Greatest Midlife Crisis

The greatest mass extinction of all time some 252 million years ago was a time of extreme global warming, when warm-climate soils spread to high latitudes due to high atmospheric carbon dioxide. This fatal atmospheric perturbation due to volcanism was reversed as carbon dioxide was consumed by soil formation.

Life on this planet was almost extinguished some 252 million years ago at the boundary between the Permian and ensuing Triassic periods of geological history. This has long been known as the greatest of all discontinuities in the history of life, more catastrophic by far than extinction of dinosaurs and other creatures some 66 million years ago. Global compilations show that some 86–97% of all species became extinct at the Permian–Triassic boundary, and life has not been the same since. The Permian–Triassic boundary was a crisis for soils as well. For example, there are no coals, which are the burial-altered remains of Histosol paleosols, anywhere in the world for the first 6 million years of the Triassic. Swamp peats that upon burial compact into coal seams, were common in wetlands for all of the past 360 million years since the evolution of swamp ecosystems, except for this 6-million-year coal gap. Other changes in paleosols, such as unusually deep weathering at high latitudes, reveal important details of this profound ecological crisis of 252 million years ago. This was the greatest known perturbation of the global carbon cycle and the rhythms of Proserpina. How close Earth came to sterilization is best appreciated by considering the varied organisms affected by the terminal Permian mass extinction.

© The Author(s), under exclusive license to Springer Nature Switzerland AG 2022
G. J. Retallack, *Soil Grown Tall*,
https://doi.org/10.1007/978-3-030-88739-1_9

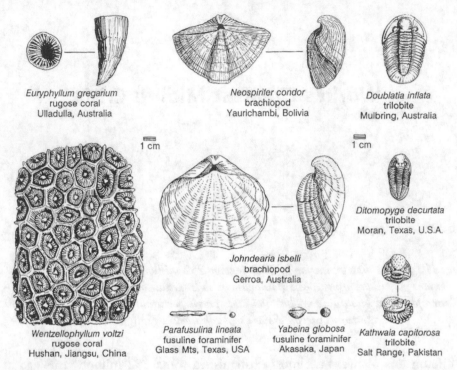

Fig. 9.1 A few of my favorite marine fossils of Permian age. Trilobites, rugose corals, and fusulinids were common for hundreds of millions of years and then became extinct at the Permian–Triassic boundary some 252 million years ago. Brachiopods were the most common and diverse bivalved shellfish before the mass extinction, but clams became more common subsequently

Before the Triassic, the most common marine shellfish were brachiopods, and the most prominent scavengers in the sea were extinct creatures called trilobites (Fig. 9.1). Today's beaches and tide pools are littered with clams and crabs, which fill similar ecological roles to brachiopods and trilobites of the past. A few brachiopods still persist, but not the extraordinary "stone butterflies," as the Chinese call them, of Permian and older times. There is nothing now like trilobites. They looked superficially like pill bugs, but had feathery external gills below each of their many walking legs. Trilobites were very successful creatures in their time, the dominant creatures of the sea bottom for millions of years after their appearance some 522 million years ago. Like other arthropods such as crabs and insects, trilobites had a jointed carapace that usually fell apart after death. A fully articulated specimen of a trilobite is a great trophy for a fossil collector and a joy to behold. Something went seriously wrong to wipe them from the face of the Earth.

Even more striking than the loss of individual species was destruction of tropical reef ecosystems at the Permian–Triassic boundary. Tropical reefs are complex biological structures of that can be seen from space. Coral reefs of the latest Permian were exuberant in beauty and biological diversity of their fossils, which can be collected in the famous reef limestones of west Texas, the Indonesian Island of Timor, and the Chinese province of Sichuan. Instead of bleached flakes of green-algal fragments and bubble-like shells of tiny foraminifera that form the most common sand grains in lagoons of Australia's Great Barrier Reef today, reefs of the latest Permian were flanked by sands of extinct coralline algae that fell apart into bead-like grains, and fusuline foraminifera that looked like stony rice grains. Instead of thickets of hexacorals, latest Permian reefs had extinct tabulate and rugose corals, which differ from modern hexacorals in having a different pattern of internal partitions. Coral reefs formed enormous masses of limestone in Permian and older rocks, but disappeared without a trace at the end of the Permian. No reefs and no corals are known from rocks dating to the first 8 million years of the Triassic. When corals reappeared in the Middle Triassic period, they were hexacorals, which are thought to have evolved from unskeletonized creatures like sea anemones, independent of Permian corals. Such profound disruptions to the fossil record of sea life have long been used to divide Earth history into a Paleozoic ("ancient life") and Mesozoic ("middle life") era. Earth's middle age, or Mesozoic, was again punctuated by the Cretaceous-Tertiary extinctions to bring on the Cenozoic or "recent life". The three great eras of geological time are not flights of human fancy, but reflect long periods of evolutionary growth between short times of crisis.

The Permian–Triassic extinctions in the sea are nowhere better displayed than in Zhejiang Province, southeastern China. An unbroken sequence of rocks across the Permian–Triassic boundary is well exposed in large quarries on the ridge running southeast of the mining village of Meishan (Fig. 9.2, Color Photo 9.1). This has become an evocative and exotic name to western geologists, but in Chinese it means merely "Coal Hill," in reference to late Permian coals nearby. The warm tropical seas of the latest Permian were floored with lime mud hosting a great variety of marine fossils: fusulinid foraminifera, brachiopods, corals, and ammonoids. Fossil shellfish are scattered in the compact dark gray limestone low in the Meishan quarries, but each fossil collected tends to be a different species from the last, so great is their diversity. Each Permian limestone bed is 4–6 in. (10–15 cm) thick, and some beds were intensely burrowed by marine worms. They appear to be deposits of shallow tropical lagoons occasionally churned by tropical typhoons.

Fig. 9.2 The Permian–Triassic boundary is between flagstones of the Changsingh Limestone and thinly-bedded shales of the Yinkeng Formation in quarry D, southeast of Meishan, Zhejiang Province, southeastern China. The concrete sign notes in English and Chinese that this is a standard reference section for latest Permian rocks in China. Much of the rock visible here in 1996 was subsequently covered by landscaping after this quarry was declared in 2001 the international standard section for the Permian–Triassic boundary

The end of this shallow-marine, tropical ecosystem is marked in the quarry walls by a clay band some 3 in. thick. This "zone of death" is in stark contrast to the fossiliferous limestones below of latest Permian age, and thick-shelled fossils and limestones of Middle Triassic age, higher in the quarries of Meishan. A distinctive bed of white claystone grades up to black shale, containing very few fossils. The shales contain small paper clams and flattened shells of ammonoids. All are small, with thin and insubstantial shells. Despite their local abundance, there is seldom more than one species on a single slab of rock. Both the paper-clams and the ammonoids were swimming and sea-bottom creatures. Fossils of burrowing creatures are rare, and fossil burrows are absent in these well-bedded gray shales. Few kinds of creatures were living on the bottom, and none dug into the sea floor. The unbalanced fossil fauna of the black shales is similar to current plague of zebra mussels (*Driessena*) in the polluted Great Lakes of North America. Ecosystems pushed to the edge by human exploitation may respond with unbalanced boom and bust cycles of population explosions. A part of the fascination of the Permian–Triassic

Color Photo 9.1 Permian–Triassic boundary extinction (252 million years) between thick Changsingh Limestone and gray shales of Yinkeng Formation, near Meishan, Zhejiang Province, China. These marine rocks are gray to white in color and well bedded

life crisis is that it was a marked reduction in species diversity comparable to extinctions of species with human modification of nature.

This great schism in the development of life on Earth is seen on land as well, from extinctions of reptiles, insects, and plants. This crisis came well before the rise of dinosaurs and mammals. Reptiles of the latest Permian are unfamiliar because they seldom appear in books and movies of dinosaurs and other prehistoric life. Reptiles of the latest Permian were diverse. Some were sleek long-limbed predators, others were stout-limbed vegetarians, or flat-headed burrowers. Some of these creatures were as large as modern rhinos, others as small as ground squirrels.

Their diversity was due in part to a variety of ecological roles, but can also be related to their geographic spread. The late Permian reptile faunas of South Africa, China and Russia were very different from one another. All this changed in the Earliest Triassic, when only a handful of species are known. The most abundant of these earliest Triassic reptiles was a creature called *Lystrosaurus* (Fig. 9.3). It is an unprepossessing looking animal that never fails to remind me of the extraterrestrial in Steven Spielberg's movie "*ET*." It had large eyes and a long drooping nose. *Lystrosaurus*, however, walked on all fours, had a stubby tail, no teeth other than two tusks, and

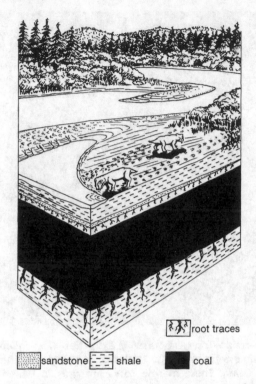

Fig. 9.3 Reconstruction of *Lystrosaurus* represented by fossil trackways in the earliest Triassic siltstones overlying the latest Permian coal in the underground mines south of Sydney, Australia (reprinted from the journal Alcheringa, 1996, with permission of the Australasian Association of Palaeontologists and Taylor and Francis)

a beak-like mouth. Adult animals of various species of *Lystrosaurus* ranged from the size of terriers to great danes. They ate low growing plants and decaying plant matter like many modern tortoises. *Lystrosaurus* is found in earliest Triassic rocks of South Africa, Russia, Antarctica, India, China, and perhaps also Australia. It was not until later in the Triassic that a greater diversity of land animals evolved, including mammals and dinosaurs. By then, there were again regionally distinct reptilian faunas on land, in contrast to cosmopolitan *Lystrosaurus*.

Most archaic insects disappeared during the Permian–Triassic life crisis. Archaic Permian insects had similar larval, nymphal and adult forms, unlike the grub, pupa and adult of most living insects. Many Permian insects were also similar to dragonflies in having paired wings that could not be folded back onto the body. A few archaic insects such as stoneflies and mayflies survived, but as in the modern world, Triassic insects were mainly beetles, flies, and grasshoppers. Few insects of any kind are known from early Triassic rocks. By late Triassic time, there were termites, wasps, moths, and other familiar insects. Here too is the familiar pattern of a few survivors of extinction giving rise to a great variety of forms tens of millions of years later.

Among land plants, turnover was similarly dramatic. In latest Permian rocks of the southern continents of South America, South Africa, India, Australia, and Antarctica the dominant fossil plant was a swampland tree called *Glossopteris*. Its name is Latin for "tongue fern," in reference to the shape of its leaves. The leaves all look so similar, that it is difficult to tell one species of leaf from another. Australian gum trees (*Eucalyptus*) have comparably uniform leaves, wood, and flowers, but some 600 species can be recognized from gum nuts. The fossilized seed structures of *Glossopteris* show great diversity, indicating that there were many different kinds of *Glossopteris* plant to the last moment of their time on earth near the end of the Permian. In latest Permian swamplands of Siberia, preserved as the great coal basin around Novokuznetsk, the dominant plants had strap-like leaves with ribs like corduroy, which is a suitable memnonic for their scientific name, *Cordaites*. The fossilized seed-structures of these plants also reveal a diversity unsuspected from their leaves. These seed-structures also disappeared at the end of the Permian. In China the latest Permian rocks deposited in swampy lowlands have preserved a variety of fossil plants: cordaites, ferns, and large leaves that look a little like those of the tobacco plant *Nicotiana*. The fossil net-veined leaves are unrelated to tobacco, and are called *Gigantonoclea* ("giant closed vessel"). This is yet another extinct seed plant that disappeared at the Permian–Triassic boundary.

Late Permian Chinese vegetation was exceptionally diverse and at low latitudes, unlike less diverse high latitude floras of *Glossopteris* and *Cordaites*. Similarly today, the jungles of Indonesia and Brazil include a riot of species compared with the beech forests of Chile and the oak-hickory forests of North America. Except for the names and a few other changes, we would have felt at home in Permian vegetation.

These varied forests of the latest Permian were all extinguished at the Permian–Triassic boundary. Passing of the forests is recorded by remarkable abundance of microscopic fungal spores and filaments, which have been extracted using strong acids from rocks at the Permian–Triassic boundary all around the world. It is as if there was a worldwide rotting of dead and dying forests. Other common fossils with the fungi are like aquatic algae of the order Zygnematales, indicating exceptionally widespread lakes of low biological diversity. The fungal-algal spike stands in marked contrast to the diverse pollen and spores liberated by the same acid-etching procedures from Late Permian and Early Triassic rocks.

Some of the most common plants in early Triassic rocks are small herbaceous forms called quillworts (*Isoetes*). These deceptively grass-like plants still thrive in nutrient-poor mountain lakes. Unlike grasses, they are spore-bearing plants. Similar forms are found worldwide during the Early Triassic, and may have been fodder for *Lystrosaurus* which has a comparable cosmopolitan distribution. Cone-bearing evergreen trees were also common after the crisis. These conifers share with quillworts an ability to thrive under low-nutrient conditions. By Middle Triassic time, another group of plants, the extinct seed ferns such as *Lepidopteris* and *Dicroidium*, had become more prominent. After decimation at the Permian–Triassic boundary, and some 6 million years of early Triassic impoverished and cosmopolitan vegetation, both species diversity and provincialism increased during middle Triassic time. This same pattern of decimation and slow recovery is repeated in plants, corals, shellfish, insects, and reptiles. The pattern also is found in the sequences of paleosols across the Permian–Triassic boundary in Antarctica, as I discovered in 1996.

Although 1996 was an *annus horribilis* for the British royal family, with the final divorce of Charles and Diana, for me it was an *annus mirabilis* of unprecedented travel and discovery. Within that one year I was able to make extended visits to both teeming metropolises of China and the splendid isolation of Antarctica. My visit to China was in part a pilgrimage to the Meishan section, now recognized as one of the most complete and informative records of events in the sea across the Permian–Triassic boundary. Graphite Peak in Antarctica, on the other hand, has one of the best known sequences of paleosols across the Permian–Triassic transition (Fig. 9.4). We camped for a southern summer month perched on a moraine at the foot of Graphite Peak, overlooking the Beardmore Glacier, with no humans other than our party of four for a hundred mile radius. Graphite Peak is about as different as one can imagine from the bicycle-clogged sidewalks and hot, humid smog of south China in mid-summer.

For many years it was difficult to be certain about the exact level of the Permian–Triassic boundary in non-marine sequences like those of Antarctica, because they lack the definitive marine fossils found in the Meishan quarries. Fortunately, there is now a geochemical method for tracing this dramatic ecological catastrophe in sediments deposited both in the sea and on land. This technique involves measurement of carbon isotopes. In normally functioning ecosystems, the isotopes of carbon in organic matter and carbonate are maintained within a limited range of proportions of the common isotope (^{12}C) to the heavy isotope (^{13}C). These proportions are regulated by a variety of processes involving the relative importance of photosynthetic production, consumption, decay and burial of carbon. At Meishan, in China, and

earliest Triassic   coal   underclay with *Vertebraria* roots   coal
Fremouw Formation boundary breccia   latest Permian
                                              Buckley Formation

Fig. 9.4 A thin claystone breccia bed, that appears to be the product of unusually intense soil erosion, marks the base of the Triassic and overlies the last Permian coal on Graphite Peak, central Transantarctic Mountains, Antarctica. *Vertebraria* is the distinctive chambered fossil root of extinct glossopterid trees. The view is south toward the South Pole, which is only 300 miles (500 km) from here

many other places as well, the carbon-isotopic composition of fossil organic matter, of the carbonate of fossil sea shells, and of the limestone itself, became suddenly and dramatically unbalanced with an excess of the lighter isotope (^{12}C) 7 in. (17 cm) below the Permian–Triassic boundary as officially defined by the appearance of a microfossil (*Hindeodus parvus*). The initial very light carbon lasted for thousands of years, before rebounding toward normal values, and then going very light again at the Permian–Triassic boundary. But normal isotopic ratios like those of the Late Permian were not attained until millions of years passed. Using chemical methods for separating carbon isotopes from fossil plant debris of Antarctica, my student at the University of Oregon, Evelyn Krull, found both unusual isotopic lightenings in several paleosol samples, and so could pinpoint the Permian–Triassic boundary in Antarctica to within a few inches. As in numerous non-marine Australian, Indian, South African, and Madagascan sequences studied in the same way (Color Photo 9.2), the great extinction proved to be above the last coal seam and the Permian–Triassic boundary about 164 ft. (50 m) higher in the sequence. Events at the boundary thus curtailed a pattern of peat

Color Photo 9.2 Permian–Triassic boundary extinction (252 million years) in gray shale with Pawa paleosols (Entisols) between Bada (thin Aridisols) and Kuta (thick Aridisols) paleosols in Palingkloof Member of Balfour Formation, near Bethulie, South Africa

deposition in Histosols that had persisted in lowlands of the southern continents for millions of years. The distinctive chambered roots and Histosols of *Glossopteris* were gone, and in their place was a new suite of soils and landscapes.

Latest Permian paleosols at Graphite Peak reveal the immediate aftermath of the great mass extinction. Latest Permian to Early Triassic paleosols lacked coal, but were gray colored and deeply weathered Inceptisols and Entisols, as would be expected in a humid lowland setting like that which existed in the latest Permian. One big difference from older paleosols of the Permian is the greater thickness and higher clay content of the postapocalyptic paleosols (Fig. 9.5, Color Photo 9.3). This is not merely due to a difference in the time over which they formed. The Permian coals are thick and probably took tens to hundreds of thousands of years to accumulate, judging from what we know about the accumulation of swamp woodland peats today. In contrast, the latest Permian and Triassic paleosols have bedding planes persisting between the root traces and clay skins, and represent soil formation over only a few thousand years for each paleosol. Despite this,

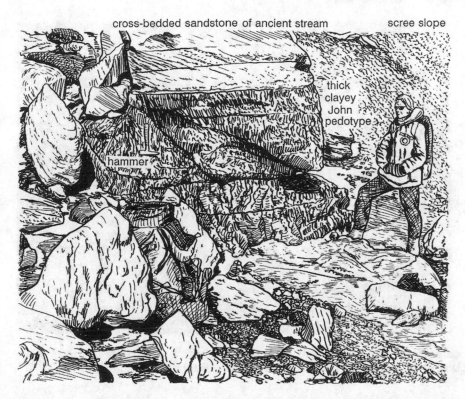

cross-bedded sandstone of ancient stream   scree slope

thick
clayey
John
pedotype

hammer

Fig. 9.5 This thick gray, clayey paleosol in early Triassic rocks of the Allan Hills, Antarctica, is unusually well developed considering the high latitude of this area, within the Antarctic Circle at that time, and is evidence of a postapocalyptic greenhouse

post-extinction paleosols are more deeply weathered chemically than the pre-extinction paleosols. Post extinction paleosols lost more calcium, magnesium, sodium and potassium over their short time of development than pre-extinction paleosols. Antarctic Permian paleosols are like those now found in frigid Canadian swamps, whereas latest Permian and Triassic paleosols are more like soils now forming in the greater warmth and rain of Pennsylvania and New York. The pre-extinction paleosols are similar to those we would expect from the reconstructed high paleolatitude of Graphite Peak at the time of the Permian–Triassic boundary, but the post-extinction paleosols are evidence of anomalously warm climate for such a high paleolatitude. This climatic anomaly was a greenhouse spike near the Permian–Triassic boundary, followed by three more greenhouse spikes into the first 5 million years of the Triassic.

Not only is the idea of a postapocalyptic greenhouse supported by evidence of paleosols, but the abrupt change in carbon isotopes on land and sea is

Color Photo 9.3 Early Triassic (Smithian, 249 million years) John paleosol (gleyed Ultisol), in Feather Conglomerate, Allen Hills, Victoria Land, Antarctica

best explained as due to an increased global atmospheric load of carbon dioxide and methane. The carbon isotopic composition of organic matter in Antarctica and India at the extinction level and again at the boundary level shows dramatic depletion in heavy carbon (^{13}C) so extraordinarily light that it could only come from large amounts of methane. Methanogenic bacteria have a strong preference for the light stable isotope of carbon (^{12}C rather than ^{13}C). These bacteria produce a depletion of –110 to –37, and typically –60 parts per thousand of the heavy isotope, on a scale relative to zero for a marine mollusc shell chosen as a standard. Volcanic intrusions into coal also produce isotopically light methane in large amounts. In contrast, the average isotopic composition of limestone and oceanic dissolved bicarbonate is near zero. The carbon of organisms is typically depleted from –14 to –35 parts per thousand on this scale of measurement, and even the oxidation of all organisms on earth could not shift the isotopic composition of organic matter as much as is observed at the Permian–Triassic boundary. Somehow there was release of large amounts of isotopically light methane from high latitude peats and permafrost, or from intruded sediments of shallow marine continental shelves or older coal measures. Methane would not have lasted long in the atmosphere because it reacts with oxygen to produce carbon dioxide within

7–24 years. The lack of coal in all known early Triassic rocks worldwide may also be a related phenomenon, because peatlands are important sites where carbon is buried out of reach of the atmosphere, and where methane is generated. Both methane and plant matter were oxidized on a large scale to carbon dioxide in the early Triassic atmosphere. The global coal gap and carbon isotopic shifts indicate disruption of the carbon cycle beyond the local confines of a few Australian and Antarctic lowlands. In this abruptly induced carbon dioxide greenhouse, a new flora and fauna found opportunity. Fragmentary remains of conifers and seed ferns and skeletons of *Lystrosaurus* found just above the Permian–Triassic boundary at Graphite Peak are similar to impoverished fossil assemblages in Early Triassic rocks worldwide. This strange cosmopolitan biota may reflect an atmosphere unusually rich in carbon dioxide during and following the mass extinction.

Remarkable successions of sedimentary rocks like those at Meishan and Graphite Peak demonstrate that the Permian–Triassic boundary was a time of far-reaching global change, but why? The list of plausible culprits goes on and on. Some theories kill off marine life by means of unusually low salt concentrations, by lack of oxygen, or by large amounts of carbon dioxide welling up from the deep ocean. These various propositions have some supporting, though circumstantial, evidence, but fail to explain the pattern of extinctions on land. Death can come from abrupt changes in salt, oxygen, carbon dioxide, and volcanic ash, but what could drive these mechanisms to such world-encompassing extremes? This diversity of culprits reminds me of a wonderful scene at the end of the classic movie "Casablanca" when Rick Blaine (played by Humphrey Bogart) has just shot a Nazi major who tried to stand in the way of escape of his former lover and her husband. The local Prefect of Police who witnessed the shooting, tells his men who come running up shortly afterwards "Major Strasse has been shot. Round up the usual suspects." The near-total obliteration of life on Earth near the Permian–Triassic boundary calls for more than just the usual suspects.

By the time our team from the University of Oregon was working in Antarctica the end-Cretaceous scenario of *T. rex* and the crater of doom was less controversial, so we made a determined effort to look for evidence of impact as an explanation for the Late Permian extinction. We dissolved the boundary bed in strong hydrofluoric acid until only quartz grains remained, and found that some of them were etched like quartz grains shattered by meteorite impact. Tiny carbonaceous chondrite meteorites were found in my samples by Asish Basu of the University of Rochester, and Stein Jacobson of Harvard University found associated cosmic spherules. Luann Becker then of the University of California at Santa Barbara found fullerenes in some of my samples and other rock samples from Permian–Triassic boundary beds worldwide. Fullerenes, affectionately known as buckyballs after the geometrically similar geodesic dome of Buckminster Fuller, are spherical 60–120 carbon atom cages, and known to be produced by wildfires and large extraterrestrial impacts. Frank Kyte of the University of California in Los Angeles found enrichments of the rare element iridium in the same samples, but they were at levels of parts per trillion rather than parts per million as at the Cretaceous-Tertiary boundary.

This all seemed promising for our efforts to demonstrate Late Permian giant impact, until we started to think carefully about the numbers. Meteoritic debris is constantly raining down on earth creating spectacular displays of shooting stars and occasional small impacts. Isotopic swings across the Permian–Triassic boundary are much larger and lasted longer than for the Cretaceous-Tertiary boundary, as would be expected from the relative magnitude of extinction, which was much greater for the Late Permian. However, shocked quartz grains are smaller and less common in the Late Permian than at the Cretaceous-Tertiary boundary. The amount of iridium is also much less in the Late Permian than at the Cretaceous-Tertiary boundary. Both are evidence of smaller Late Permian impact or impacts. The Antarctic beds do not show such marked chemical leaching as the Cretaceous-Tertiary impact beds at Brownie Butte, Montana. We had gone all the way to Antarctica with significant government funding, and had found some of what we were seeking, but the numbers did not add up for a theory linking severity of mass extinction to magnitude of impact. One of the hardest things for a scientist to do is admit a favorite hypothesis is wrong. But that is what we did, and then moved on to more promising explanations.

The theory that ended up sticking had to do with methane from thermogenic cracking of coals by the enormous flood basalts of Siberia from the Late Permian to Early Triassic. Our first clue was carbon isotope values so extremely light that at first we did not believe them. We tried several laboratories, but were eventually convinced when numerous analyses by Hope Jahren and Bill Hagopian of the University of Hawaii returned the same values, and tests for total carbon and different rock matrices supported their veracity. When the ratio of heavy (^{13}C) over light (^{12}C) carbon is less than −40 per thousand only one culprit remains, and that is biogenic or thermogenic methane (CH_4). This severely unbalanced isotopic ratio is due to the overwhelming preference of methanogens and other microbes for the light isotope of carbon. There are huge reserves of methane, or natural gas, locked as clathrate in ice of permafrost and ice of shallow marine sediments of high latitudes. Microbial methane is also associated with black shales and coals, and can be mobilized by fracking to release as natural gas. Our Antarctic analyses of Late Permian carbon isotopes, together with a global network of strong negative perturbation, showed that the isotopic excursion required about 2000 gigatons (1 gigaton is 10^{15} g) of carbon in methane with a typical isotopic depletion of -60 per thousand. That is more carbon than in all living organisms and soil humus today, suddenly lofted into the air, where the unusual mix would be fixed by plants and incorporated with plant debris in the rocks. Getting this amount of methane from clathrate ices, would be difficult because methane clathrate is very stable in its intergranular underground reservoirs, and requires extraordinary measures to release. Furthermore, oil company estimates of this potential fuel show that no single clathrate reservoir hit by an asteroid today could give the globally observed Late Permian isotopic perturbation. Perhaps there were larger reservoirs in the past, but that seems unlikely given global increases in productivity since the Permian. A more likely source of the observed Late Permian methane pollution of the atmosphere is the hot feeder dikes of large igneous intrusions such as the Siberian Traps, which erupted through Tunguska Coal Measures. Coals would have been cracked to methane on either side of the feeder dikes then vented to the air in unparalleled volume. The Siberian Traps have long been known as the largest flood basalts known in the last 500 million years, and radiometric dating of them showed that they peaked at the same time as the volcanic ashes associated with the mass extinction at Meishan. All the numbers aligned to implicate atmospheric pollution by flood basalt eruption in the greatest of mass extinctions. Large basaltic eruptions of the Deccan Traps of India accompany the Cretaceous-Tertiary extinctions, and other eruptions also coincide with other mass extinctions at the Triassic-Jurassic boundary, and during the Late Devonian, and Late Ordovician. We went to Antarctica thinking that large impacts may have brought on mass extinctions, with large igneous provinces a minor contributor or perhaps a result of impact. Now it seems that volcanic pollution is a major player in most mass extinctions, and impact had a role in only one. Science often amazes, but occasionally also surprises.

Methane in the atmosphere could kill in a variety of ways, by reacting with atmospheric nitrogen to form formaldehyde, by igniting, or by anaesthetizing, and asphyxiating. There would have to be more than 5% by volume of methane for significant formaldehyde generation or atmospheric burning, and more than 0.5% by volume is a standard for mine safety, beyond which humans experience difficulty breathing. The amount of methane released at the Permian–Triassic boundary can be calculated from the magnitude of the carbon isotopic shift in organic matter. The answer is about 0.03% by volume of the atmosphere. This is not enough poison or narcotic on a global basis. The principal effect of methane at the Permian–Triassic boundary was probably as a greenhouse gas. Methane is fifty times as effective as carbon dioxide in reducing the loss of solar energy to space and so encouraging global warming. Methane also oxidizes to the greenhouse gas carbon dioxide within about 7–24 years. Global warming also has the effect of evaporating water, and water vapor is another potent greenhouse gas. Was the terminal-Permian dying a product of postapocalyptic greenhouse?

The idea of death from an unusually severe methane, carbon dioxide, and water vapor greenhouse resonates with a variety of observations that have been used to support other theories about the Permian–Triassic extinctions. In the ocean, excess carbon dioxide has already been advanced as an explanation for the preferential extinction of corals and brachiopods with heavy shells, poor respiratory and circulatory systems and low basal metabolic rate. A high concentration of dissolved carbon dioxide also explains unusual crystals of calcite that appear to have precipitated on Early Triassic ocean floors. The black shales, preservation of complete undecayed fish, and lack of burrows in earliest Triassic marine rocks have been taken as evidence of oceanic stagnation. Today, stagnant ponds and seas are starved of oxygen by high levels of biological productivity that produces carbon rich sediments, but in contrast, earliest Triassic shales of lakes and seas have very low organic carbon contents. The methane release hypothesis gives an alternative mechanism for depleting oxygen in the ocean.

High carbon dioxide, methane and water vapor also explain many peculiarities of early Triassic soils and plants. Early Triassic paleosols of Antarctica are chemically weathered like soils of Pennsylvania and New York, unlike the Late Permian paleosols which indicate conditions like those of Canadian boreal swamps. Some of the Early Triassic soils are comparable with Ultisols, which are now found no further north or south than 58° latitude, whereas this part of Antarctica was at least at 70°S during the Early Triassic. Fossil Ultisol indications of greenhouse paleoclimate in polar regions are confirmed by studies of fossil leaves of a seed fern called *Lepidopteris*, which was a common tropical plant of Permian times. By earliest Triassic time *Lepidopteris* had extended its range southward into the Antarctic Circle, which then included Madagascar and southeastern Australia.

Fossil *Lepidopteris* leaves also give direct evidence of high carbon dioxide levels. Some fossil leaf specimens retain a black carbon film of the original leaf, including the leathery outer coating or cuticle of the plant, which preserves microscopic outlines of the epidermal cells and breathing pores, or stomates. Latest Permian *Lepidopteris* fossil leaves from Australia, India and Russia have unusually low numbers of stomates on their leaves. Modern *Ginkgo* leaves, which are distantly related plants, have been reducing their stomatal abundance over the past century as carbon dioxide levels rise from fossil fuel consumption. Using studies of this change in modern *Ginkgo* as a calibration, the earliest Triassic greenhouse had carbon dioxide an order of magnitude more than in the atmosphere today.

Increased carbon dioxide and reduced oxygen of the earliest Triassic would not have been a serious problem for plant leaves, which evidently adjusted their intake, but would have created difficulties for plant roots, which respire like us. In wetland soils, only marginally aerated at the best of times, lack of oxygen would have killed trees at the roots. Tree dieback, fungal rotting, rising water tables, then a regrowth of ferns and other herbaceous plants are revealed by the fossil record of coal, leaves, pollen and spores across the Permian–Triassic boundary. Deforestation and soil erosion worsened rising atmospheric carbon dioxide, and is indicated by widespread thin layers of clayey soil clods, now preserved as claystone breccia in many Permian–Triassic boundary sections. Microscopic examination of those clayey clods shows clay films and other microstructures characteristic of clayey parts of soils usually broken down to suspended clay and rolling grains by transport in water.

More evidence for landscape destabilization is apparent from a change from meandering Permian streams to Triassic braided streams of the sort that drain clearcut forests and deserts. Meandering stream deposits include a mix of narrow paleochannel sandstones and strongly-developed paleosols, whereas braided stream deposits are mainly sandstone, with weakly developed paleosols. Methane outburst and its oxidation to carbon dioxide in the atmosphere not only explains how plants were killed, but the observed cascade of landscape destabilization.

Loss of forest food and shelter was difficult for reptiles, and only generalized small detritivores like *Lystrosaurus* survived, but reptiles faced other difficulties from hydrocarbon pollution of the atmosphere. According to the computer modeling of Yale geochemist Robert Berner, the large amount of methane in the terminal Permian atmosphere indicated by carbon isotopes consumed much oxygen in its conversion to carbon dioxide, and together with ongoing extinctions and volcanic eruptions, lowered atmospheric oxygen levels to only 12% by volume, much lower than today's 21%. This reduced atmospheric oxygenation would be comparable to living in mountains 14,000 ft. (4500 m) high, where mountaineers now run the risk of nausea, headache and worse. In such low oxygen atmosphere, death can come from pulmonary edema, which like pneumonia, fills the lungs with fluid and stimulates coughing of a frothy blood-specked sputum. Is this how many Permian reptiles died? Circumstantial support for this idea comes from a variety of adaptations found in survivors of the atmospheric crisis, which have a rudimentary secondary palate separating the buccal from nasal cavities and also a relatively short face for easier passage of air into the lungs. *Lystrosaurus* and other latest Permian to Early Triassic reptiles also have barrel chests and a reduced number and size of belly compared with chest ribs, unlike their more weasel-like ancestors. These adaptations are evolutionary steps toward the mammalian condition of a long secondary palate and a muscular diaphragm behind the ribs of the chest. Flat faces and barrel chests also are found in Quechua Indians of the Andes and Nepali sherpas today, who are adapted to lower oxygen levels of high altitude. *Lystrosaurus* also is found in burrows. Such close quarters would have selected for respiratory adaptations which enabled it to survive late Permian air pollution.

This theory of late Permian death by impact released methane can be called the Ragnarok scenario. In ancient Norse mythology, Ragnarok is the twilight of the gods, bringing our current cycle of life to an end and ushering in a new age of harmony. The myth begins with a triple winter, associated with earthquakes, sea level change and the heavens torn asunder, as would be expected from massive eruptions and volcanic dust shroud. The myth goes on to explain the ensuing cold and hunger that forces the gods to assemble on the plain of battle and fight to the death. Then the giant Surtur having killed Freya, the goddess of spring, consigns the world to fire and flames in a way comparable with a methane greenhouse. The Ragnarok scenario is a disturbing vision of biotic vulnerability to impact or volcanic disturbance. Life on Earth has been affected by large extraterrestrial impacts and massive basalt eruptions, which both totally annihilate their local surroundings. Global effects come from atmospheric injection of carbon dioxide and methane on unprecedented scale giving a spike of greenhouse warming and eliciting respiratory difficulties. There are clear implications for current oxidation of oil and coal polluting our atmosphere and creating warming by the greenhouse effect. The Ragnarok scenario is of interest as an example from the past of atmospheric carbon dioxide and methane levels of devastating proportions. As far as we know, that was as bad as atmospheric pollution ever has been. We need to know more about mass extinction events of the past because they reveal limits of life on Earth. The cycles of the Proserpina Principle, designed to roll with the punches, were almost undone at the end of the Permian.

THE GREAT DYING

World's greatest midlife crisis was when

Lost species were more than nine of each ten.

When most kinds of Permian forest tree died

Rot and fungal spores spread worldwide.

Insects and reptiles died out with the greenery,

And survivors adapted to different scenery

Of weeds in warm, impoverished ground.

The sea was choked with decay all around,

And old-fashioned seashells became extinct.

The next ten million years were distinct

Lacking reefs of coral or swamps of peat,

So bad was the time life met near defeat.

10

Roots of Trees

As trees evolved from small early land plants some 370 million years ago, they reinvigorated soil formation to produce novel kinds of soils, Alfisols, Ultisols, Spodosols and Histosols. Enhanced and deeper biological weathering depleted atmospheric carbon dioxide to produce widespread glaciation.

Trees and forests have a fossil record extending back some 370 million years to the middle of the Devonian geological period. In addition to fossil logs and leaves, paleosols are also a record of ancient forests. Several kinds of soils today are distinctive of forests: the fertile clayey Alfisols of Indiana's oak-hickory forests, infertile clayey Ultisols of Oregon's Douglas fir forests, infertile sandy Spodosols of Vermont's spruce woodlands, and thick peaty Histosols of Louisiana's cypress swamps. The evolution of these kinds of forests and their soils had profound effects on landscapes and paleoclimate. Trees and these distinctive soils were an entirely new element in long term rhythms of the carbon cycle. When and why did they evolve?

Life with and without forests was a pressing political issue in my home state of Oregon, as the lumber industry retooled for lower yields of second growth forest, and lumberjacks and sawmills lost employment. Eugene and the University of Oregon are at the southern end of the Willamette Valley, a broad north–south trough between the Coast Range and the snowy peaks of the Cascade Mountains in the western part of the state. Cascade volcanoes provide ash and lava for potentially fertile soils, but the mountain ramparts intercept moist winds off the Pacific Ocean. Although our summers are as warm and dry as in many Mediterranean lands, the wet season used to start

© The Author(s), under exclusive license to Springer Nature
Switzerland AG 2022
G. J. Retallack, *Soil Grown Tall*,
https://doi.org/10.1007/978-3-030-88739-1_10

during October and last until May. Abundant rain weathers the soil deeply, and this combined with snowy winters and a hot dry growing season is tough on deciduous trees. Thick, red clayey Ultisols support tall forests of conifers draped in ferns and floored with mosses. The few remaining cathedral forests are awe-inspiring places.

Such primeval forest is becoming increasingly difficult to find as logging has become more widespread and effective. Clear-cutting of forests has left large areas of stumps, erosional gullies, and streams choked with silt and muddied with clay. Forest fires induced by climate change are also decimating our forests. In the late 1990's, the so-called "hundred year flood" has become annual, as clear-cut logging has proceeded and weather has turned from a relatively dry decade toward one of increased rain and less snow. The work of our lumberjacks makes it easy to see the importance of forests in holding the land together. Left alone, trees intercept the rain so that it does not erode the soil directly. Instead, rain-water is passed from leaf to leaf and down the trunk onto mossy soil rich in litter and a variety of soil animals. The slowed passage of water from air to ground feeds clear perennial streams that are shaded from the sun, and deep and cool enough to support salmon and trout. The great sponge of forest has played an important role in preventing floods and landslides after rain.

In stark contrast is the desert of eastern Oregon, which lies in the rain shadow of the Cascade peaks. The open spaces, dry clear air, and smell of sage and juniper have a charm of their own. The thin soils and abundant rocky outcrops of the desert expose numerous colorful paleosols, which are an archive of 50 million years of climate change from subtropical humid forests to the barren high desert of fragrant sagebrush today. Unlike the thick soils and forests that cover rocks of western Oregon, desert Aridisols of eastern Oregon are thin, silty and salty. Much rock and bare earth is exposed between the scattered sagebrush and juniper trees. Windstorms can whip the parched silty soil into a blinding dust storm. Cloudbursts can turn the soil into a torrent of mud that rapidly fills dry creek beds. The dusty campfire circle at Hancock Field Station, a perennial favorite of my student geological excursions in central Oregon, was once overwhelmed by a sudden red rush of turbid water. Life is only marginally in control of this barren land, where wind and water rule with sudden and unpredictable ferocity.

Deserts give some idea of soils and vegetation before forests evolved. There is also evidence in the form of fossil plants and soils of the Devonian geological period, some 359–419 million years ago. Only herbaceous fossil plants have been found in earliest Devonian strata, but fossil tree trunks are common by the middle of the period. Some Middle Devonian fossil

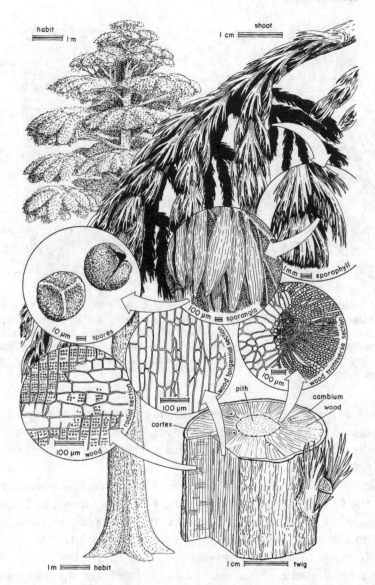

Fig. 10.1 Reconstruction of a middle to late Devonian tree based on fossil leaves called *Archaeopteris macilenta* and fossil wood called *Callixylon newberryi* from upper New York state (reprinted from Retallack 2019, Soils of the Past, 1990, with permission of the J. Wiley and Sons)

tree trunks are 5 ft. (1.6 m) in diameter. The fossil trees are extinct, with a curious combination of spore-reproduction and conifer-like woody anatomy unlike anything alive today (Fig. 10.1). Their superficially fernlike leaves have been called *Archaeopteris* and their massive trunks were called *Callixylon*, long

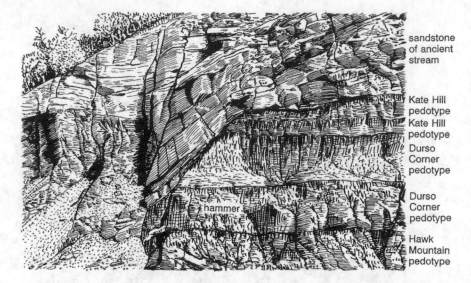

sandstone
of ancient
stream

Kate Hill
pedotype
Kate Hill
pedotype
Durso
Corner
pedotype

Durso
Corner
pedotype

Hawk
Mountain
pedotype

Fig. 10.2 Middle Devonian (379 million year old) red paleosols of three different kinds (pedotypes) truncated to the left by a steep sided ancient stream deposit, in a road cut on the Catskill Front near East Windham, upper New York state. The steep erosional truncation of well developed paleosols is the former cut bank of a meandering stream, which became more common as forests began to alter patterns of rainfall and runoff

before it was realized that these were different parts of the same plant. A few fossil trees do not necessarily a forest make, but forest ecosystems are clear from paleosols associated with fossil plants where they are best known, in upper New York state.

In the rural part of New York, 150 miles northwest of Manhattan, green hills of the Catskill Mountains are dotted with sleepy towns and laced with miles of stone fences. The hard flagstones and rocky summits are sandstones formed in Devonian rivers draining westward from a formerly magnificent mountain range of 370 million years ago, now eroded back to the hills of western Massachusetts. In the gray-green sandstones of these ancient rivers and the dark gray shales of associated ponds are locally abundant remains of fossil plants, including the earliest fossil trees. Also found interbedded with these drab sediments are numerous massive red-and-green mottled beds, representing the ancient soils that once supported trees (Fig. 10.2, Color Photo 10.1). These paleosols have many features of forested soils, such as Alfisols. The paleosols are thick (usually 3–4 ft. or about 1 m), with a subsurface clayey (Bt) horizon, as is typical under forests where clay washes deep into soil along stout, tapering root channels. Calcium-rich nodules deep within the paleosols and persistent feldspar grains indicate that these soils were fertile

Color Photo 10.1 Late Devonian (366 million years) Hyner (thick Aridisol) and Bucktail (thin Aridisol) paleosols in the Duncannon Member, Catskill Formation, near Newport, Pennsylvania

with plant nutrients, calcium, magnesium, potassium and sodium. The paleosols also are riddled with abundant stout woody root traces. Sometimes the root traces include fossilized wood. More often though, the woody root traces are thick striated channels filled with red clay or rimmed with green-gray mottles, produced during the decay of the root in stagnant groundwater shortly after burial of the paleosol. Fossil root traces are abundant and extensive in the paleosols, as in forest soils of continuous canopy. The degree of weathering and of root-burrow reworking, and depth of the calcium rich nodules within the paleosols are all evidence of forests of warm, subhumid paleoclimates.

Not all the fossil soils in Devonian rocks of the Catskills supported forests. Entisol paleosols, black with carbon, and crammed full of the remains of herbaceous clubmosses (*Colpodexylon*), probably supported marsh vegetation. Inceptisol paleosols mottled green and gray with spreading tree roots of early fernlike plants (*Wattieza*), represent former swamp woodland and intertidal woodlands comparable with modern mangroves. Other Aridisol paleosols are red and were probably well drained, like the forested Alfisol paleosols, but with only small root traces, shallow calcium-rich nodules, and common burrows, are like soils of dry shrublands today. Plants are very uncommon in these paleosols and include leafless green twiggy shrubs (*Rellimia*). Taken together, this suite of paleosols is evidence of an ancient coastal plain with a mosaic of marsh, swamp, shrubland, and forest. Although not so luxuriant or diverse in its plant species as savannas and monsoon forests on comparable soils today, Middle Devonian forests were much more substantial and complex communities of plants than herbaceous vegetation of earlier times.

Not only was Devonian vegetation of New York varied in space, it was also varied in time, cued to a series of transient greenhouse spikes. The forests and their deep-calcic paleosols migrated westward with each of the greenhouse spikes, displacing the shallow-calcic dry shrubland paleosols. These transient greenhouse spikes have long been known from black shales formed in the ocean at the same time as the forested deep-calcic paleosols. The black shales are rich in organic matter from when shallow marine waters lost oxygen, and many sea creatures died. This is a new source of natural gas energy in the region, released by fracking the black shales in deep wells. Each of these anoxic events was hard on shellfish in the ocean, and coincides with turnover in Devonian the fossil faunas. Two of these black shales in Germany are called the upper and lower Kellwasser event, which formed at the same time as a pair of deep-calcic forested paleosols in New York. These twin greenhouse spikes stand out as a time of global mass extinction comparable in magnitude with the Cretaceous extirpation of dinosaurs. They also coincide in time with the Viluy basalts of Siberia and Pripyat-Dniepr-Donets basalts of Ukraine. Like the Late Cretaceous Deccan Traps and Late Permian Siberian Traps, these massive eruptions may have polluted the atmosphere with large amounts of thermogenic methane and carbon dioxide. This stimulated globally warmer and more humid conditions thus facilitating forest expansion into aridlands of New York and elsewhere. This larger biomass and deeper weathering in turn caused drawdown of atmospheric carbon dioxide, so that the greenhouse spikes increased and declined rapidly, punctuating a long term trend of global cooling from forest tree and soil evolution.

Long term effects of newly evolving forest communities can be seen in river channel deposits associated with Early and Middle Devonian paleosols. Early Devonian and geologically more ancient river deposits mostly have shallow, wide sandstones with broad scours marking the eroded edges of the channel. These were streams like those of deserts, choked with sand-bars so that the stream of water was divided into a braid of shallow rivulets. Braided streams persist in sparsely vegetated, disturbed and sediment-choked watersheds today, but in Devonian deposits of the Catskills, meandering stream paleochannels became common for the first time in geological history. In these ancient river channels, one side of the sandstone bed is cut off steeply, whereas the other side is feathered out into laterally equivalent siltstones and shales. One side represented the steep undercut bank in the outside of a river bend. The other side was a point bar building into the inside of the river bend. Such meandering streams form on floodplains of low gradient and clayey texture, but are also promoted by forest vegetation in several ways. Trees resist flood scouring by their baffling action on flood waters. Trees also make a clayey floodplain by promoting deeper and more thorough weathering of sand-sized mineral grains. The continuous and active transpiration of water by trees also nucleates clouds, mist and rain to even out annual fluctuations in soil moisture. There is an increased abundance of meandering stream deposits from Middle to Late Devonian geological time as forested Alfisol paleosols appeared and became widespread in the rock record. This observation supports the idea that forests gained unprecedented control over physical forces of landscape erosion and deposition.

Middle and Late Devonian changes in vegetation and streams also had consequences for the evolution of tetrapods from fish. A classic evolutionary narrative from Harvard's Al Romer, is that the limbs of fish were selected by escape from desert ponds in the seasonally dry landscapes of the Catskill Delta of New York. This plucky lobefin fish jauntily setting forth into the unknown is widely celebrated in cartoons and advertising. There are indeed seasonally dry, shallow-calcic, aridland paleosols in the Catskills, as Romer noted when developing his theory. Sadly these paleosols do not host either bones or footprints of early tetrapods, which are all associated with the calcareous forest soils of woodlands. The best known of these sites is near the hamlet of Hyner in northwest Pennsylvania, where a long road cut has yielded fragmentary bones of at least three species of early tetrapods. These bones are fragmentary as is usual for remains mixed into soils, but they are similar enough to other known skeletons of early tetrapods to support a surprising fact about the earliest amphibians. Unlike frogs and newts, Devonian tetrapods had elongate bodies, vertical fishlike tails, and very short legs. They were aquatic creatures adapted with their stumpy limbs for negotiating vegetation choked waters of oxbow lakes, and the shallow ponded floodwaters of woodlands. They did however have narrow necks like frogs and newts, and so could tilt their heads to feed in shallow waters in a way impossible for fish. The new habitats and new flooding regimes created by newly evolved woodlands, also selected for evolution of the earliest tetrapods. Shrinking desert ponds on the other hand were fatal to fish (Fig. 10.3).

Other kinds of forested soils also appeared in the rock record not long after Middle Devonian paleosols of fertile forests. The oldest known coals

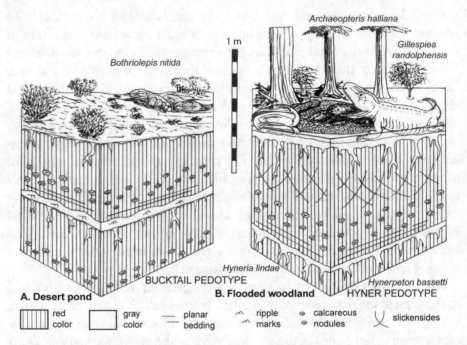

Fig. 10.3 Reconstructed aridland (**A**) and woodland (**B**) soils at the Late Devonian tetrapod locality of Hyner, Pennsylvania. The woodland hypothesis for tetrapod origins proposes evolution of limbs in flooded woodlands and vegetation choked lakes, rather than shrinking desert ponds (reprinted with permission of the Journal of Geology, 2011, and University of Chicago Press)

or Histosols formed by woody plants are found in Late Devonian rocks of 360 million years ago in road cuts around the small West Virginia town of Elkins. Some early non-woody land plants of the Early Devonian formed Histosols, but not the thick, carbon-rich coals made possible by swampland trees. Coal-bearing paleosols of ancient swamps are more common in the succeeding Carboniferous geological period, so named because of these thick carbon reservoirs. Histosols remained common in every succeeding geological period (Fig. 10.4). Mining of these great reserves of coal fueled the Industrial Revolution of eighteenth century Great Britain and North America. The thick black seams of coal were formed in swamps like those of Okefenokee Swamp in Georgia. In these wooded wetlands, the trees that die do not decay but accumulate in stagnant water as thick brown layers of fibrous peat. With burial and time, the peat is purged of mobile elements such as oxygen and hydrogen, and becomes a black seam of coal, rich in carbon. Most of these coal seams represent ancient swamp soils or Histosols.

Histosols are unusually difficult soils for plants, because they are acidic from both decay products of plants and antibiotics produced by plants to

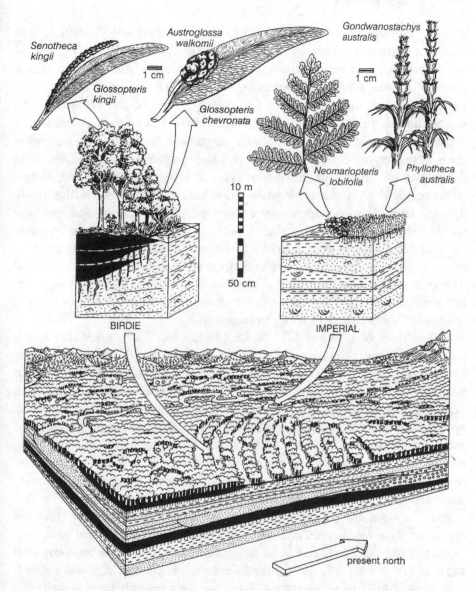

Fig. 10.4 Reconstructed soils and plants of latest Permian deciduous swamp wood-lands of *Glossopteris* in the Sydney Basin, southeastern Australia. The ridges in the swamp are string bogs of a kind now found in cold temperate climates. Drawdown of the Devonian carbon dioxide greenhouse by trees and their soils ushered in the Permo-Carboniferous glaciation and these cool temperate swamplands (reprinted from the Bulletin of the Geological Society of America, 1999, with permission of the Geological Society of America)

mitigate animal and fungal attack. They also are very low in essential plant nutrients derived from the weathering of mineral grains. Histosols also tend to be oxygen poor, which discourages root growth. Unlike leaves which photosynthesize using carbon dioxide, roots require oxygen to respire. Low oxygenation was also a problem for swamp plants of the Devonian and Carboniferous, as can be seen from the tabular form of their roots, which spread out laterally over the moderately oxygenated surface and do not penetrate deep, oxygen-starved portions of their paleosols. Considering these difficulties for growth in peaty substrates, it is not surprising that forests of waterlogged ground appear geologically later than those of fertile, freely drained land. Tabular root systems for coping with soil oxygen shortage would not have taken long to evolve, because roots do not grow in stagnant parts of soils. Adaptations to acid and infertility, such as plant production of phenolic and chelating compounds to more aggressively liberate mineral nutrients, were more complex evolutionary solutions. Perhaps long-term biochemical evolution explains why trees adapted to peaty soils so many millions of years later than to well drained soils.

Other kinds of soils are difficult for plants even if well drained. Oxisols for example, are thick, deeply weathered, clayey soils, which have very low reserves of the nutrient elements, calcium, magnesium, potassium and sodium. Oxisols are known well back into Precambrian time more than 1660 million years ago, but evidence of forests on such infertile substrates appear even later in geological time than forests of waterlogged ground. The most ancient forested Oxisol known to me is Carboniferous in age, some 305 million years old. In the rolling Ozark foothills of Missouri are many open pits that exploit red, deeply weathered claystones in sink holes of limestone and dolostone. Some of these red claystones contain robust woody root traces. The vegetation of these ancient Oxisols can only be inferred from the nature of these fossil roots and soils because plant fossils are not preserved in such highly oxidized paleosols or sediments. Comparable modern soils support rain forest: tall, multistoried forest of warm, wet climatic regions. This general kind of vegetation is likely for the Carboniferous as well, but the plants of these Carboniferous rain forests were completely unrelated to flowering plants that dominate modern rain forests. Carboniferous rain forest plants were probably broad-leaved seed ferns, as known from nearby northwestern Illinois within shales deposited in ravines of an ancient landscape comparable to the tower karst landscapes of south China today.

Another difficult soil is the clean quartz sand of Spodosols. Their quartz is mainly elemental silicon and oxygen, and they also have a subsurface horizon rich in iron and humus, but the essential cationic nutrients, calcium, magnesium, sodium and potassium, are scarce. The oldest known example of a Spodosol is a Carboniferous paleosol found in quarry near Bristol, England. Stout root traces in this sandy paleosol indicate former forest vegetation. Yet another difficult soil is the deeply weathered clay of Ultisols, which have much aluminum, as well as silicon and iron, but also are low in cationic plant nutrients. The oldest known Ultisol is a paleosol of Carboniferous age from the Lykens Valley of northeastern Pennsylvania. Carboniferous Oxisols, Spodosols and Ultisols indicate that forests had by this time extended their influence even further than fertile floodplains and waterlogged bottomlands to nutrient-poor uplands. The stabilizing effect of forests was thus extended to an even greater fraction of the landscape than during the Late Devonian.

Early forests did more than just hold ground against erosion. Like grasslands of the past 33 million years, these early forest soils may have been agents of global change in the carbon cycle. The burgeoning biomass of forests sequestered a significant portion of the world's inventory of carbon. Much of it was in the form of lignin, the essential component of wood. Lignin is a triumph of biochemical evolution, a molecule that changed the world. Lignin is slow to decay and is eaten by very few creatures, and so an important adaptation to plants under attack by herbivorous insects. Devonian insects were not equipped to deal with wood on the scale of later evolving termites and dinosaurs. Trees reduced atmospheric carbon dioxide to lignin and other plant carbon compounds by photosynthesis and other biochemical pathways. Some of this carbon went into peaty paleosols and was buried as coal below the surface of the Earth for hundreds of millions of years until exploitation by humans. Even well drained soils were sprinkled with chips and other residues of wood. The forest soils were eroded to sediment richer in carbon than sediment of earlier geological times. Deeply weathered soils promoted by the stabilizing effect of the early forests also consumed carbon by weathering in unprecedented amounts. Carbonic acid in these soils liberated nutrients for forest growth, with a loss of carbon exported in groundwater as bicarbonate to the sea. The source of the weathering solutions of carbonic acid was carbon dioxide from the atmosphere and respiration of soil animals and microbes in a zone of weathering deepened by large tree roots. Thus forests and their soils consumed atmospheric carbon dioxide at rates unprecedented in geological history.

There were limits to forest's effectiveness in this grand task of atmospheric engineering because replacement of atmospheric carbon dioxide with oxygen liberated by photosynthesis has the effect of making forests easier to burn. Forest fires restore carbon dioxide to the air, but also create charcoal, which is slow to decay and release its carbon. Charcoal first turns up in paleosols of Early Devonian age. Early forest ecosystems created a new equilibrium of atmospheric oxygen and carbon dioxide. With forests the world changed from greenhouse conditions of earlier times to an oxygen-rich and carbon-dioxide-poor atmosphere comparable to that of the present.

These theoretical atmospheric effects of the first forests and their soils are confirmed by several studies indicating a decline in atmospheric carbon dioxide during the Devonian appearance of forest stumps, paleosols, and charcoal. One line of evidence comes from study of carbon isotopic composition of organic matter and calcium carbonate in paleosols. Plants have a strong preference for the light stable isotope of carbon (^{12}C rather than ^{13}C), and so have isotopically lighter carbon than found in the atmospheric carbon dioxide which is used to manufacture their leaves and wood. The isotopic composition of carbon dioxide in the soil is derived in part from fungal decay and other consumption of soil organic matter from plants. It is soil respired carbon dioxide mixed with atmospheric carbon dioxide diffusing into the soil that determines the isotopic composition of carbon in calcium carbonate nodules in the soil. Carbon dioxide diffuses further into the soil when carbon dioxide levels of the atmosphere are high, thus shifting isotopic values of carbonate nodules to heavy values. Conversely, low carbon dioxide in the atmosphere is easily overwhelmed by respired carbon dioxide in soil pores, and the soil carbonate nodules become isotopically lighter. Measurement in a mass spectrometer of the carbon isotopic composition of organic matter and calcium carbonate in paleosols shows strongly declining values of atmospheric carbon dioxide through Devonian time. A second line of evidence comes from study of fossil land plants, or more particularly the small pores, or stomates, through which carbon dioxide diffuses into the plant, and oxygen and water vapor are transpired. Plants have fewer stomates when atmospheric carbon dioxide levels are high, and more stomates when carbon dioxide is scarce. This line of study also shows declining atmospheric carbon dioxide through the Devonian.

Carbon dioxide is a greenhouse gas because it impedes reflection of the sun's light energy back into space. The Devonian decline in atmospheric carbon dioxide went hand in hand with global cooling and the extensive glaciation of the following Carboniferous and Permian periods. Great ice sheets spread at that time over the southern continents of South America, South Africa, Australia, and Antarctica, then united in the supercontinent of Gondwana. In the Hunter Valley region of southeastern Australia, for example, Permo-Carboniferous glaciers left scratched glacial pavements in bedrock, and thick beds of bouldery till. Glacial deposits also include pale-osols cracked and banded by ground ice in a manner comparable with modern Gelisols. Late Permian lake deposits of shale in the sea cliffs near Newcastle, at the mouth of the Hunter River, in Australia, have a distinctive seasonal banding called varves, reflecting slow accumulation of clay under winter ice, followed by silt of spring runoff. Fossil leaves of *Glossopteris* asso-ciated with these varved shales are arranged in thin rafts within the base of the winter layer, as if the plants were seasonally deciduous. This early example of deciduous habit is also indicated by healed scars from shedding at the basal attachment of the fossil leaves and from the asymmetry of growth rings in the fossil wood of these trees. Burrows of archaic herbivorous reptiles, perhaps for hibernation, also date back to the Permian, and are particularly well known in rocks of this age from the Karoo Desert of South Africa. Decid-uous trees, hibernating animals and glacial landscapes are familiar features of the modern world, and date back to well before the dinosaurs. Deciduous trees, cool climate herbivory, and ice sheet expansion were further limits on photosynthesis and atmospheric oxygenation, in addition to fire and char-coal. These complex biotic feedback mechanisms that now play such a role in planetary atmospheric regulation can be traced at least as far back as the Permo-Carboniferous ice age.

With forests came new regimes of environmental regulators. Deeper soils, greater weathering, massive tree trunks, and stream containment all promoted greater environmental stability, carbon dioxide consumption, and climatic cooling. On the other hand, wildfires, deciduous leaves, and growth of polar and alpine glaciers added new elements of instability. A new equilibrium was struck in which forests flourished in a variety of humid climate soils, such as Alfisols and Histosols, which formed for the first time in geological history during the Devonian period. Older kinds of soils of highly seasonal climates (Vertisols), very dry climates (Aridisols), and disturbed ground (Entisols) were not entirely supplanted by new forests and their soils. Forests were added to increasingly heterogeneous vegetation and soilscapes. Like the global spread of grasses and angiosperms, lignin and trees were evolutionary inventions that came to have global consequences by cooling the planet.

FOREST HOME

Oregon has old-growth forested spaces,

And deserts rimmed by bare cliff faces.

Sage and greasewood dot dusty plains.

Floods of mud follow cloudburst rains.

Life's difficult in desert, chancy, hard

Unlike cool, ferny forests which guard

From blazing sun and swollen stream.

Raindrops pass from leaves that gleam

To ferns of spongy soils that sieve

It clean so salmon and trout can live.

Ages ago, the pattern was sown.

Forests make a home for their own.

11

Mighty Millipedes

Paleosols provide evidence for the nature of plants and animals on land before the evolution of vascular land plants. Millipedes and liverworts with us still, played an important role on land more than 440 million years ago.

The public image of scientists has changed little over the years, despite increasing scientific and technological literacy. Mad scientists are now portrayed in movies as more greedy and antisocial than ever, consumed with the potential power of their discoveries. Few shed a tear at the violent demise of the brilliant genetic engineer, the quirky computer technician, or the garrulous bearded paleontologist in the "Jurassic Park" movies. In truth, scientists, and their discoveries are as varied as humanity. We often do not know what we will find, nor where our detective work will lead. Many scientific discoveries come from luck and collaboration. As you read this there are scientists somewhere flying to scientific conferences, drinking beer in draughty exhibition halls, or lecturing in overheated dim rooms. When geologists go to a meeting, the lure of the field is also strong. Getting there is most of the expense, so why not get a feel for local rocks and fossils? Pilgrims visiting a holy site find a similar beauty and reverence in connecting with the physical manifestations of their inner life.

Sometimes these pilgrimages uncover something new, as in 1983 during a post-conference trip in central Pennsylvania, where I toured fossil sites for trilobites and sea shells remaining from prolific marine life of the late Ordovician, some 450 million years ago. Pennsylvania also has red sedimentary rocks from ancient rivers and coastal plains of the same great geological

© The Author(s), under exclusive license to Springer Nature Switzerland AG 2022
G. J. Retallack, *Soil Grown Tall*,
https://doi.org/10.1007/978-3-030-88739-1_11

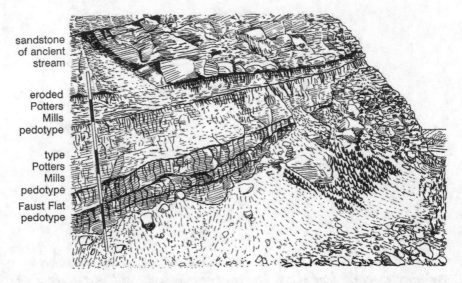

sandstone
of ancient
stream

eroded
Potters
Mills
pedotype

type
Potters
Mills
pedotype

Faust Flat
pedotype

Fig. 11.1 Late Ordovician (444 million year old) paleosols in red beds of a long road cut east of Potters Mills, Pennsylvania. The paleosols are red, clayey, and burrowed horizons within these sandy beds

age, brought to my attention by Richard Beerbower during conversations at geological meetings. Richard's career at the State University of New York in Binghampton has placed him well for the study of rock of this age throughout the Appalachian Mountains. His tip was good, because the red beds contained what turned out to be evidence of the earliest animals then known on dry land. This kind of discovery cannot be made in books or in a computer, but only in the great outdoors. My discovery was exciting because landscapes at that time in the geological past had previously been envisaged as barren as the surface of Mars. I had found something that noone suspected was lost: evidence that millipedes were among the earliest animals on land.

The critical evidence came from a long road cut east of the small village of Potters Mills in central Pennsylvania, where Late Ordovician red beds are riddled with tubular burrows of animals (Fig. 11.1, Color Photo 11.1). Broadly similar burrows have long been known from marine rocks of Ordovician age, so these burrows had been taken as evidence of a shallow sea in this area. My inspection of the red beds uncovered numerous observations that they were not formed in the sea or a lake but as ancient soils. Even in the field could be seen the telltale clay enrichment, clay skins, and small calcium-rich nodules that are characteristic of soils. Furthermore, some nodules had grown across the burrows, whereas other nodules were cut by the ancient burrowing animal. Thus, the burrows formed at the same time as the nodules and the

Color Photo 11.1 Late Ordovician (444 million years) Potters Mills paleosols (Aridisols) and Faust Flat paleosols (sandy Entisols) in the Junata Formation, near Potters Mills, Pennsylvania

soils, rather than in marine sediment later exposed to soil formation, or in a soil that was later inundated by the sea.

Laboratory studies confirmed that that these rocks were soils and that these burrows were dug by soil animals. Such extraordinary claims require extraordinary evidence. The variation in chemical composition up and down the ancient profiles was very similar to that expected during weathering. Calcium, magnesium and sodium were depleted from the profiles. These elements would have been enriched at the surface of the beds if they had formed by settling through water rather than as soils. The ratio of isotopes of carbon in the carbonate nodules was very low like that of soil nodules rather than like marine limestone. Furthermore, isotopes of carbon and oxygen in the carbonate were very highly correlated with each other, as expected in soils and plants, but not in marine limestones. The orientation of clays and internal fabric of the carbonate nodules seen in thin sections of the rocks under the microscope are exactly like those of soils. The red, iron-enriched nature of the paleosols indicates that these paleosols were formerly well drained, and rusted by exposure to atmospheric oxygenation, rather than stagnant and water-logged. The shallow depth of calcium-rich nodules in the paleosols indicates an arid climate, because such nodules form deeper within soils of subhumid climates, and do not form at all in humid soils strongly leached of calcium. The burrows in these paleosols were evidence that animals were living in dry dusty soils as long ago as 444 million years.

Fossils of the animals themselves have not yet been found. Nor should they be expected in such oxidized paleosols, where they were prone to decay and dissolution. I still hold out hope that they will turn up in suitable gray shales of lakes. Nevertheless, much can be learned about these early land creatures from the nature of their burrows. The burrows were quite large for such ancient land dwellers. The burrows ranged in size from 0.1 to almost 1 in. in diameter (2–25 mm), but their size did not show a continuous distribution of diameters. Instead, a few sizes were more common than others. This is a typical growth pattern for arthropods such as shrimp and insects. Their hard external skeleton is periodically shed as the animal grows. A new skeleton is secreted once the animal pumps itself up to a slightly larger size to give more growing room. The burrows also contain back-filled clay and silt in W-shaped layers as if they were packed by a bilaterally symmetrical animal. Bilaterally symmetrical arthropods include trilobites, but their external gills would have wilted in such dry soils. Centipedes have flexible bodies and squeeze through existing soil cracks, rather than excavating burrows like the fossils. Shrimp, crabs and insects have appendages of more varied size and shape than indicated by the regular succession of backfills. By a process of elimination then, the most likely excavators of these fossil burrows were millipedes.

Millipede body fossils of Ordovician age continue to elude discovery, although Ordovician tracks like those made by millipedes have recently been found in the Lake District of Britain and in Kalbarri National Park, Western Australia. However, millipede fossils are well known in non-marine rocks of the succeeding Silurian and Devonian periods, and some poorly preserved fossils like millipedes and centipedes have been found in marine rocks as old as Cambrian, confirming a 510 million year geological antiquity of broadly comparable archaic arthropods (Fig. 11.2).

Living millipedes were a childhood fascination of mine, and my own young children also caught and played with millipedes. Such slow-moving creatures are easy to capture and appear archaic compared with the exuberant array of hopping, flying and running insects in the garden. Millipedes remind me of the horse-drawn buggies of the black-clad Amish slowly ambling past my outcrop of Ordovician paleosols in Pennsylvania, as cars and trucks zoomed by on the high-speed highway. Working on field notes at road cuts I have learned to tune out the whoosh of passing cars and trucks, but the slowly approaching clip-clop of Amish buggies always took me by surprise.

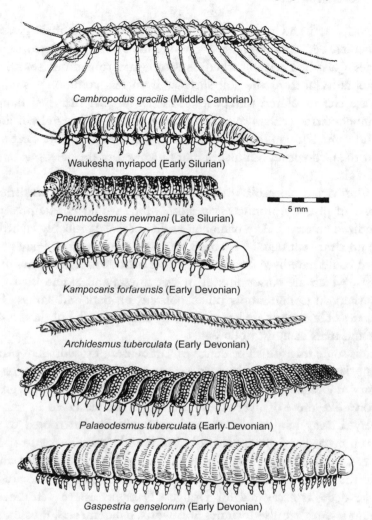

Cambropodus gracilis (Middle Cambrian)

Waukesha myriapod (Early Silurian)

Pneumodesmus newmani (Late Silurian)

5 mm

Kampecaris forfarensis (Early Devonian)

Archidesmus tuberculata (Early Devonian)

Palaeodesmus tuberculata (Early Devonian)

Gaspestria genselorum (Early Devonian)

Fig. 11.2 Reconstructions of Cambrian to Devonian fossil centipedes (*Cambropodus*) and millipedes (others) from Britain and Canada (reprinted from Retallack 2019, Soils of the Past, 1990, with permission of the J. Wiley and Sons)

Amish farms are widespread in this part of Pennsylvania, and are notable for their lack of connection to local electrical lines, and for teams of horses plowing the fields. Their religious beliefs prohibit them from participating in technological advances of the twenty first century. Millipedes similarly are anachronisms among modern soil fauna.

A variety of millipedes can be found in most suburban gardens, where they consume rotted plant debris, and sometimes live vegetables as well. In my Oregon garden, the flat-backed millipedes of leaf litter (*Harpachne*) have colorful red and yellow patches along the sides of their glossy dark brown bodies. Less commonly seen alive are the cylindrical burrowing snake-millipedes (*Julus*). Their bleached skeletons are sometimes found curled up like a ram's horn at the soil surface. Even less commonly seen at the surface are cream-colored millipedes, with an angular marginal flange and a head much narrower than their body (*Polyzonium*). These polyzoniid millipedes and some Silurian fossils from Britain (*Cowiedesmus*) are closest to my vision of the Ordovician millipedes that burrowed in the paleosols of central Pennsylvania.

The Ordovician paleosols with millipede-like burrows tell us little about the nature of primary producers on land, which would not be preserved in such oxidized paleosols. Any plants or similar herbage probably lacked roots, because no clear root traces have been found in the paleosols. Plants without roots that could have lived on these paleosols include mosses, hotnworts, and liverworts, which are common still in wet forests and alpine bogs. Mosses form feathery tufts, commonly called polsters, on bark and stones. The flat green blades of liverworts are less commonly encountered, but these also form polsters and mats in moist hollows.

There is some microfossil evidence of Ordovician liverwort-like plants. By dissolving large amounts of rock in strong acids, my late colleague at the University of Oregon, Jane Gray, has found distinctive microscopic groups of four spores like those of liverworts. The spores are preserved in gray shales but decayed away long ago in the red paleosols. She attempted to extract them from my red and green paleosols, but found none. Liverwort spores are widespread in Ordovician and Cambrian rocks of North America and elsewhere in the world. Middle Ordovician liverwort megafossils also are known from lake deposits excavated for Douglas Dam in eastern Tennessee. This fossil plant assemblage also includes hornworts, balloonworts, mosses, mycorrhizae, and lichens. This extraordinary locality is only a few miles from Dolly Parton's tourist resort Dollywood, and I named one of these fossil mosses *Dollyphyton* after her.

Judging from the red and calcareous nature of Ordovician paleosols in Pennsylvania, their ancient vegetation lived under drier conditions than familiar thalloid liverworts (*Marchantia*) crowded around waterfalls and river banks. The spores found by Jane Gray are most like those of bottle lichens (*Sphaerocarpos*), and there are other fossils similar to leafy liverworts (*Cephaloziella*), both found still in well drained soils. Communities of non-vascular plants of well-drained ground may have been widespread in Ordovician time. Small gray-green patches remaining from organic matter in the paleosols are scattered like polsters of mosses and lichens on rocks and fenceposts. Such vegetation now and in the past can be called a polsterland, which is meant as a general term comparable to woodland and grassland. Polsterland consists of mosses, liverworts and lichens and comparable rootless and rhizomeless vegetation with a patchy distribution. Since the Ordovician these communities have been supplanted in fertile well-watered soils by forest and grassland, as outlined in previous chapters. Communities entirely of non-vascular plants are found today mainly in extreme desert and frigid wastelands.

Today, liverworts, mosses, and millipedes persist within communities of vascular land plants. Mosses and liverworts are commonly the first to invade bare or disturbed ground. They are especially adept at exploiting bare rock faces. The polsters are then oases for later colonizing grasses and daisies. These in turn build the soil further by their litter and by trapping dust so that trees can take root. Mosses and liverworts are not entirely displaced, but persist in the forest floor and in crevices of bark. Despite their unsophisticated form and lack of defenses, mosses, liverworts and millipedes have persisted in modern ecosystems. Modern ecological succession of bare ground recapitulates an evolutionary sequence in which liverworts, mosses and millipedes first colonized the land.

Early land ecosystems of millipedes supported by scattered clumps of liverwort-like plants may have appeared similar to modern desert and tundra polsterlands (Fig. 11.3), but differed fundamentally from any modern community. These differences have been revealed by careful study of the isotopic composition of carbon in carbonate nodules at various levels within Ordovician paleosols. For a start all the isotopic values in the Ordovician paleosols were somewhat light, that is to say lighter than modern soils in the ^{13}C isotope of carbon compared with the ^{12}C isotope. This may indicate a lower amount of carbon dioxide in the atmosphere than now, so that isotopically heavy air diffused less deeply into the pores of the soil during the Ordovician than is the case now. Calculations using simple diffusion models proposed by Thure Cerling of the University of Utah yield amounts of Late Ordovician atmospheric carbon dioxide less than a half of the modern level. Such low levels are also seen in bubbles within Antarctic and alpine ice formed during the last Ice Age of 40,000 years ago, and indicate that the Potters Mills paleosols formed during the Late Ordovician ice age. This Hirnantian ice age has long been known from rocks exposed in the Sahara Desert of Africa, which was then at much higher latitude than now. The evidence includes glacial scratching, dropped pebbles, ridges of angular rock like tills of glaciers, and paleosols with ice cracking features like modern Gelisol soils. It is sobering to

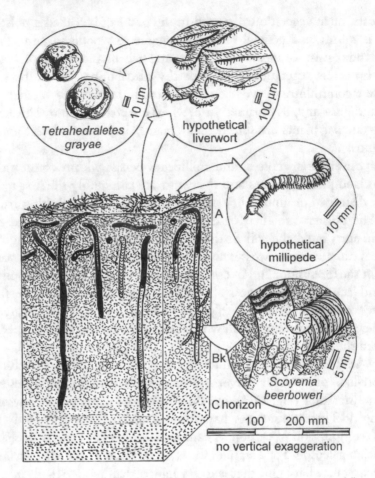

Fig. 11.3 Reconstruction of a Late Ordovician (444 million years old) soil, with burrows and vegetation, from central Pennsylvania. This early land vegetation of liverworts, browsed by herbivorous and detritivorous millipedes, can be called a polsterland (reprinted from Treatise of Geochemistry, 2007, with permission from Elsevier)

see such evidence of frigid conditions in the Sahara Desert, one of the hottest parts of the world.

In contrast, another paleosol only a little older within the Late Ordovician (445 million years old) reveals as much as 16–18 times the carbon dioxide in the preindustrial atmosphere. This thick Oxisol profile is atop the Neda Iron Formation near a village of the same name in southeastern Wisconsin. This Neda paleosol has within its iron oxides organic matter with heavy carbon isotopic values changing smoothly within the paleosol to lighter isotopic values created by soil respiration deeper within the paleosol. Unlike the Potters Mills paleosol, the Neda Oxisol is evidence of one of the steamiest greenhouse climates of geological time coinciding in time with another mass extinction of marine life in the earliest Hirnantian part of the Late Ordovician. This mass extinction was comparable in severity with the Cretaceous-Tertiary boundary extinction, and like it has a dramatic greenhouse spike. Rather than from the Late Cretaceous Deccan Traps of India, this Late Ordovician greenhouse spike may have been from thermogenic methane and carbon dioxide released from large basaltic eruptions of the Yerba Loca and Alcaparrosa Formations of the Argentinian Precordillera. The eruptions and greenhouse spike were a short interruption of long-term cooling to the Hirnantian ice age from the diversification and spread of nonvascular land plants such as mosses and liverworts. The rise of plants thus countered an earlier long-term greenhouse created by millipedes and other large consumers on land, which reset the carbon budget of soils, as envisaged also for the geologically later advent of termites and dinosaurs in the Jurassic. Although interrupted by a greenhouse spike and mass extinction, the advent of Ordovician land plants initiated cooling and ice caps in the same way as the Devonian-Carboniferous evolution of trees by increasing biomass on land and promoting deeper chemical weathering by carbonic acid than before. The evolutionary dance of producers and consumers according to the Proserpina Principle regulates long term variation in greenhouse gases and paleoclimate. Nutrient, water and gas economies of soils, by virtue of their wide extent, can have global consequences.

LAND ANIMALS

Lichens, mosses, millipedes, still

After countless ages, as always fill

In open spaces, even bare stone,

Then grasses make the soil their own.

Millipedes graze growing microbial mats,

Then are hunted by insects, lizards and rats.

The sequence mimics the millions of years

Before forest and shrubland appears.

When green plants first lived on land

Consumers were very close at hand.

Since shunted aside by active lives,

Lowly creatures manage to thrive.

12

Lichens and Till

Back before evolution of land plants and animals more than 540 million years ago, soils were similar to those of desert and alpine regions today, and like them supported microbial earth communities, including lichens.

Outback Australia is a magical land of rocks and paleosols that informs us about landscapes and life of truly remote periods of geological time. The red soil and unrelenting flatness are broken by mere stumps of mountains, worn with age into mythical shapes. The great ribbed turtle-back of Uluru (formerly Ayers Rock) has the grandeur of a Gothic cathedral. The dawn howl of dingoes coming in to drink at its springs can be as haunting as a church organ reverberating from stone columns and arches. One part of the Australian outback is of particular importance to geologists, a land of rock ridges named for one of Australia's pioneer navigators, Matthew Flinders. The distant hills of the Flinders Ranges are blue and gray above the red soil, and framed by the ghostly white bark of river gums. The landscape and soils are ancient, probably some 100 million years old. But the rocks are more ancient still. They were formed in a world before sea shells and trilobites, during the late Precambrian, 550–1000 million years ago. The story of those distant times is archived in upturned layers of rock exposed in the deep canyon of Brachina Gorge on the western margin of the Flinders Range (Fig. 12.1, Color photo 12.1). Some of this rock sequence was formed in shallow seas which accumulated gray limestones with fossils of marine algae and cyanobacteria, but many thick bands of red beds show telltale signs of paleosols: cracked, clayey and stained red with the iron-oxide pigment,

© The Author(s), under exclusive license to Springer Nature
Switzerland AG 2022
G. J. Retallack, *Soil Grown Tall*,
https://doi.org/10.1007/978-3-030-88739-1_12

Fig. 12.1 Late Precambrian paleosols (red slopes) and paleochannels (vertical cliffs) in Brachina Gorge, Flinders Ranges, South Australia

Color Photo 12.1 Ediacaran (550 million years) Yaldati paleosol (Aridisol at base), Warrutu (thick Gelisol), then Warru (Entisols) and Muru (Aridisols) paleosols in the Ediacara Member, Rawnsley Quartzite, in Brachina Gorge, South Australia

hematite. These ancient soils and associated fossils are puzzling and important evidence for Proserpinan mechanisms of global change well before the evolution of life as we know it now.

The red paleosols and the overlying flaggy sandstones are famous for a remarkable assemblage of fossils named the Ediacara biota. The Ediacara Hills, where the fossils were first discovered, are a low desert range barely visible to the northwest from the heights of the Flinders Ranges. Another term for these strange fossils is vendobiont, after the ancient Vend tribe of northeastern Russia, where comparable fossils have been found. Similar fossils of about the same geological age have now been found on every continent except Antarctica. Most vendobiont fossils are preserved as large impressions in clean white sandstone variably stained with the red ocher of weathering. From the time they were first discovered by Australian geologist Reginald Sprigg in 1947, these fossils were interpreted as remains of soft-bodied sea jellies, sea pens and worms. Because they predate the first shelly fossils and trilobites, they were imagined as ancestors to the great evolutionary radiation of sea creatures during the early Cambrian geological period. Some paleontologists still interpret them that way, but these fossils have recently become both controversial and fascinating.

I first had doubts about this interpretation many years ago, while making my own family camping and student excursion pilgrimages to the Flinders Ranges from 1967 to 1975. First of all, some of the Ediacaran creatures were huge. I saw one 3 ft. (1 m) long and still going down into the rock with a taper indicating that another third of the fossil was still covered. Richard Jenkins of the University of Adelaide has estimated that some of these fossils were almost 5 ft. (1.5 m) long. How could these enormous fossils have been ancestral to the tiny known earliest Cambrian and latest Precambrian fossil sea shells? I have seen great heaps of these small early Cambrian seashells in the phosphorite mines of Yunnan, China, and all are less than 0.1 in. (2 mm) in diameter, as they are in rocks of comparable geological age all around the world. Giant clams and squid evolved much later in geological history. The ancestors of the earliest fossil sea shells should have been smaller still, not these Ediacaran giants.

Even more puzzling was the way these supposedly soft-bodied creatures made such a deep impression on rock surfaces. Many of them were recessed several tenths of an inch (5 mm) below the surface of the slabs of sandstone that buried them. How could that be, when the fossils had been covered by a thickness of more than 4 miles (6 km) of overlying rock? Under such deep burial conditions, soft-bodied fossils should be squashed to the thickness of a sheet of paper. Some of the fossils were also peculiar in sticking up into the

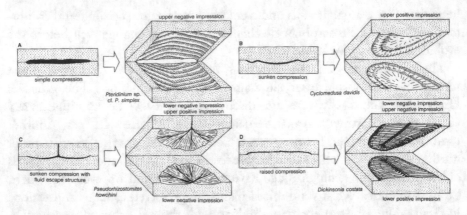

Fig. 12.2 These four different preservational styles of Ediacaran fossils are surprisingly thick and resistant to compaction considering their burial under some 4 miles (6 km) of overlying rock. Sea jellies and sea pens would have been crushed to the thickness of paper under such a weight of rock. Could these fossils have been lichens fortified by structural chitin? (reprinted from the journal Paleobiology, 1994, with permission of the Paleontological Society and Cambridge University Press)

overlying sandstone that buried them, apparently resisting its flattening effect (Fig. 12.2). In addition, the very preservation of soft-bodied fossils in quartz sandstone was a problem. Soft-bodied fossils are preserved in limestones and black shales at such famous localities as Chengjiang in China, Burgess Pass in Canada, and Solnhofen village in Germany. But how could they be preserved in sandstone composed almost entirely of quartz and leached of virtually all organic matter? It was all very puzzling and strange.

Adolf Seilacher of the Universität Tübingen in southern Germany, pointed out additional riddles presented by the Ediacaran fossils in a series of articles beginning in 1984. The supposed sea jellies had the deepest impressions of muscle masses in their centers, rather than around the periphery of the umbrellas as in living sea jellies. The supposed sea pens were solid, thick structures more like quilted bedspreads, than the open branching structures of modern sea pens. The supposed segmented worms had alternating segments on either side of the midline, rather than segmentation all the way across the body, like true segmented worms. None of the fossils had any sign of a mouth, gut or anus (Fig. 12.3). For all these reasons and more, Seilacher proposed that these famous fossils could not be ancestors of geologically younger, small animal fossils, but were rather a failed early experiment in the history of life. He explained the unique preservation of Ediacaran fossils by a unique decomposing biota at this remote time in geological history, less effective than the fungi and bacteria that have destroyed a fossil record of soft-bodied creatures in oxidized sediments ever since. Or perhaps Ediacaran fossils were

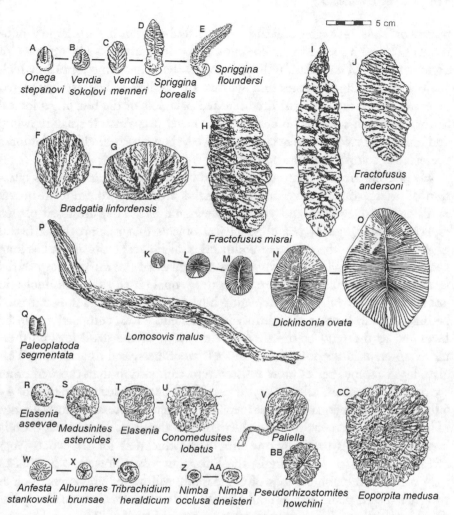

Fig. 12.3 A variety of Ediacaran or Vendobiont fossils arranged in likely growth series and all drawn to the same scale (reprinted from the journal Paleobiology, 1994, with permission of the Paleontological Society and Cambridge University Press)

preserved by an unusually thick marine microbial mat, which thrived before the evolution of snails, fungi and other enemies of microbial mats.

Seilacher's biological arguments were compelling, but the claims for evolutionary and preservational uniqueness were less so. Decomposers were clearly effective at that time because Ediacaran fossils are associated with a variety of fossilized trails. The bodies of the creeping and burrowing soft-bodied animals which made these small trails are exceptionally rare, and reported only recently as *Ikaria*. The small trails are most like those made by tiny worms or "slugs" of aggregated slime molds, and quite unlike the large vendo-bionts. The white sandstones containing the Ediacaran fossils are poor in organic matter of any kind, unlike the black shales and limestones in which geologically younger leeches and worms are fossilized. The associated red paleosols are cracked and oxidized like red paleosols of the rest of geological history in which soft-bodied creatures are never preserved. If small crawling and burrowing creatures were largely decayed, then why were the vendobionts preserved if comparably soft-bodied?

My paleobotanical background inspired me to look further into this question of preservation. Unskeletonized remains are locally common in quartz sandstones of Devonian and younger geological age. They are fossil plants, especially fossil logs. Logs are cylindrical objects that can provide a useful gauge for assessing the degree of compaction of Ediacaran fossils. It has long been known that cylindrical fossil logs are squashed flat during deep burial in rocks. Because burial pressure at depth is equal on all sides, fossil logs do not spread out laterally with increasing burial depth, but are simply flattened because they are weaker than lithifying sediment. Thus, compaction can be measured as the ratio of the thickness to width of the fossil log. Further-more, different kinds of logs vary in their woodiness, and thus density. This gives logs of some species more resistance to compaction than those of other species. Measurement of a variety of fossil logs once buried to a depth of 4–5 miles (6–8 km) in quartz sandstone, like the Ediacaran fossils, shows a range of compaction from about one thirtieth the thickness for densely-woody logs of seed plants to about one fiftieth the thickness for less-woody logs of fossil clubmosses. Surprisingly, the Ediacaran fossils proved to be thicker than compacted clubmoss logs of comparable width, and almost as thick as compacted seed-plant logs. It could be that Ediacaran fossils were also cylin-drical originally and squashed flat during burial like logs. That explanation does not work because there are Ediacaran fossils a tenth of an inch (2 mm) thick and an inch (25 mm) or more wide in beds of sandstone only a quarter of an inch (6 mm) thick. If the original fossil was spherical and thus an inch (25 mm) or so tall, such a thin layer of sediment would have banked up against the fossil, rather than covering it. If the Ediacaran fossils were thinner than wide, this would require much greater compaction resistance than club-moss logs. The inescapable conclusion is that Ediacaran fossils were tougher than wood!

Ediacaran fossils were much more resistant to compaction than soft-bodied sea jellies or worms. Nobody has ever found a trace of shell or bone on Ediacaran fossils, and the fossils are not crazed with cracks like crushed shells and bones. They must have had some kind of tough compound, comparably strong to the lignin of wood. An obvious possibility is chitin, as in fungi. Could the Ediacaran fossils have been lichens? At last I had a workable hypothesis, and the beginning of two wonderful decades of controversy and discovery.

Lichens are consortia of algae and fungi, of cyanobacteria and fungi, or of cyanobacteria and actinobacteria. Actinobacteria need not be considered in this context because they lack compaction-resistant chitin of fungi. Cyanobacteria and green unicellular algae are familiar as green pond scum formed from these photosynthetic microbes. The fungal partner provides shelter and structure. Lichens are often touted as an example of mutual cooperation between organisms, or symbiosis. Another view of the fungal partner now popular is that they imprison their photosynthetic partners, because the liberated photosynthetic partner grows larger and faster on its own. Lichens come in a staggering array of sizes and shapes, including radiating crusts, leafy blades and furrowed mats. Their component microbes may have a Precambrian fossil record, but it is sparse and of uneven quality. The oldest well preserved fossil lichen is early Devonian, some 150 million years younger than these South Australian fossils.

Sherlock Holmes made famous the technique of solving a mystery no matter how improbable by eliminating the impossible, but the improbable idea that Ediacaran fossils were lichens was at first difficult even for me to swallow. For six years I introduced the idea as an entertaining proposition for cocktail parties at geological conferences. Most of my colleagues laughed, regardless of how much they had to drink. There were many objections to this heretical idea, but in the end none had much substance. Science unlike religion or philosophy does not prove propositions. Rather it proceeds by testing propositions through experiment and observation, rejecting those that do not withstand scrutiny. The lichen hypothesis proved difficult to falsify. Falsification works best if it is your own hypothesis you have disproved, like my change of mind concerning asteroid impact for the Late Permian mass extinction in Antarctica. The Ediacaran lichen hypothesis is still controversial, but remains to be disproven.

"Those permineralized specimens are just red algal fragments or decayed animals!" my critics warned me concerning the best preserved Ediacaran fossils from China and Namibia. Hans Pflug of the Universität Giessen in Germany had described some Ediacaran (550 million years old) fossils from the southern African country of Namibia, and showed clearly that they did not have the cell structure of animals. These were permineralized in silica so that details of their cell structure were preserved, as best known from

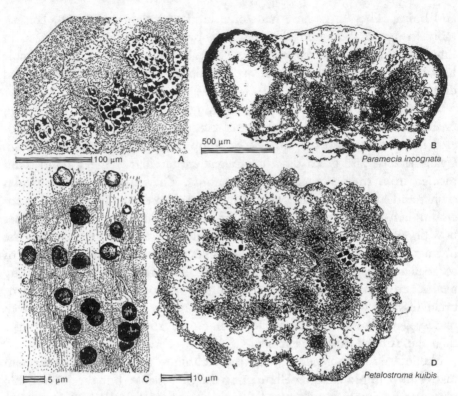

Fig. 12.4 Microstructures of Precambrian Vendobionts from Namibia (below) and China (above), showing tubes (fungal mycelia) and dark spots (pigmented photosynthetic algae or cyanobacteria), similar to the microstructures of lichens (reprinted from the journal Paleobiology, 1994, with permission from the Paleontological Society and University of Cambridge Press)

fossil wood. Under the microscope there are abundant fine tubes forming a loose meshwork between dark, round, spots. No animal or plant has tissues like this, but the mesh is very similar to fungal threads and the dark spots are like pigmented algae or cyanobacteria (Fig. 12.4). The fossils examined by Pflug were cup-shaped (*Ernietta*) and quilted (*Petalostroma*) like many Ediacaran fossils, with good exterior outlines, which would be odd for a carcass completely decayed by fungi. Furthermore, the threads are less dense on the inside than the outside of the fossil, which has most of the dark spots, as also seen in microscopic cross sections of a lichen thallus. In contrast, fungal mycelial decay within an animal should emanate from patchy centers of infection. Comparable microstructures of tubes and spots have also been found in fossils of Ediacaran age (600 million years old) permineralized by estuarine phosphorite in mines of Weng'an, Guizhou Province,

China (Fig. 12.4). Some of these were later interpreted as red algal fragments, but they lack the near-cubic cells and fourfold clusters of cells found in genuine fossil and modern red algae. Instead they have filamentous structure, with a distinct outer rind dotted with heavily pigmented cells more like a lichen thallus. Another comparable fragment with comparable filaments and dark spherical cells from Weng'an was later accepted as a lichen, probably a mucoromycotinan or glomeromycotan fungus and cyanobacterial partner, by expert paleomycologist Thomas Taylor of the University of Kansas. A shortcoming of this likely Ediacaran lichen is that it is a tiny fragment in a petrographic thin section, and it is unknown whether it had the external form of vendobionts. These fossils are important search images, and give encouragement that vendobionts may one day be better understood at a cellular level.

"But they look just like sea jellies and sea pens, so they must be marine," my cocktail party critics rejoined between sips of wine. The rocks containing the fossils in the Flinders Ranges and Ediacara Hills presented immediate problems for that interpretation. The fossils were in sandstones and siltstones with mud cracks and other features of the kind found on beaches, tidal flats, and soils, as noted by Australian geologist Sir Douglas Mawson, who was the academic advisor for Reg Sprigg at the University of Adelaide. With Sprigg advocating marine and Mawson terrestrial, a compromise explanation was that the fossils were thrown up on the beach by storms. I well remember as a teenage surfer on the beaches of southeastern Australia, that long wracks of seaweed and swarms of sea jellies on the beach were a smelly nuisance after a big storm. But the Ediacaran fossils were never found thrown together in a long tangle parallel to the shore. Nor were they all of about the same size like stranded sea jellies. Instead they varied enormously in size, with size related changes like those of growth stages. The elongate, worm-like forms, for example, had more segments in the longer specimens. The circular sea jellies-like forms had more concentric rings on specimens of larger diameter. In addition, they were well spaced on the slabs. I have never seen two overlapping Ediacaran fossils in the hundreds of slabs I have inspected in the field, and in the collections of the University of Adelaide and the South Australian Museum. But some overlaps have subsequently been documented in tall rooted forms superficially like sea pens (*Charniodiscus*). Rather than moving about like animals, the vendobiont fossils lived on the sediment where they are preserved. By this view their overlap was prevented by competition between ground-hugging individuals for living space, as in lichens.

"I've never seen lichens in and around the sea! What are you talking about?" cocktail party companions complained, reluctant to abandon the argument that they were marine. Until recently, I had also overlooked the abundance of lichens at the sea shore. The rocky shores of Oregon and Washington, for example, are a mosaic of colorful patches in the intertidal zone: black felts of *Verrucaria*, orange pustular smears of *Caloplaca*, irregular yellow fuzz of *Xanthoria*, and gray, radially folded discs of *Physcia*. There are also small, green lichen buttons and films that live entirely in the sea. After years of beach combing and tide-pool exploration, it was a thrill to discover something new in a place I thought I knew well. Marine biologists have known and written about marine lichens for years, but they do not often find their way into natural history guides. It takes microscopes, chemical laboratories, and culture vessels to really appreciate these simple, and unprepossessing creatures.

My own suspicions echoed the published opinion of Douglas Mawson based on red color and loess-like textures that the rocks with Ediacaran vendobionts in South Australia were paleosols. His arguments had not prevailed, and nor would my publication of a few field observations from my youth. Thus began a decade of field work and nostalgia in South Australia beginning in 2003. My recalcitrant cocktail party colleagues would only be persuaded by overwhelming evidence, and probably not even then. Or perhaps my quest would disprove my own hypothesis, as happened to me in Antarctica. Field indications were very promising indeed, because many of the red bed sequences with Ediacaran fossils had diagnostic field features of paleosols: cracking and other bedding disruption down from erosional surfaces, red clay skins in cracks, and diffuse subsurface horizons of sand crystals and nodules. I also confirmed Mawson's observations of red paleosol clasts in grey sediment and finely alternating red and green beds, that the oxidation from green to red occurred during the Ediacaran, rather than in the current outcrop. A sand crystal has the form of a crystal, with regular planar faces, and is not clear inside, but studded with other mineral grains. Sand crystals are characteristic of soils, especially desert soils, and the Ediacaran examples were originally gypsum desert roses like those of Aridisols. The nodules of calcite were also shallow in the profile as found in other kinds of Aridisols. Some of the Ediacaran paleosols with vendobiont fossils were deformed in a way like frost boils found in marginal permafrost soils, or Gelisols.

All these features were documented in maps, sections and panel diagrams, but then work began in the laboratory. Thin sections of the rocks with Vendobionts confirmed the internal chambered structure noted by Seilacher (Fig. 12.5). Thin sections also confirmed the shape and size of loess grains

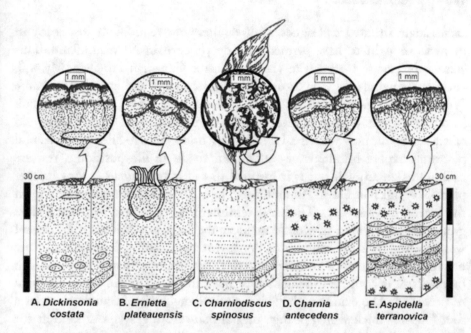

Fig. 12.5 Selected vendobionts and their paleosols from South Australia (**A**), Namibia (**B**) and Newfoundland (**C–E**), showing general internal structure inferred from thin section examination (reprinted from the journal Alcheringa, 2016, with permission from the Australasian Association of Paleontologists and Taylor and Francis)

noted by Mawson, but also the exact grain size distribution of loess. Chemical profiles of the paleosol demonstrated the characteristic depletion of calcium, magnesium, potassium, sodium, and phosphorus found in soils, but not marine sediments. Furthermore stable elements such as titanium were concentrated in the tops of the profiles in a fashion only found in soils formed by loss of weathered volume, rather than marine sediments in which titanium in heavy minerals sinks to the bottom of the bed. Isotopic values of carbon and of oxygen are also very light and highly correlated with each other as in soil calcites, and unlike marine limestones. Finally, after these results had been published and caused quite a stir, other lines of evidence emerged from the germanium content of silica cements in the Ediacaran fossils. It is high in Ediacaran holdfasts as in soils and paleosols, rather than low as in marine cherts and silica cements. Boron content of the fossils is low as in non-marine fossils, unlike high boron in marine fossils.

"Well what about the Vendobionts found in deep marine rocks? How could lichens dependent on photosynthesis survive at water depths beyond those of light penetration?" I would be asked between sips of wine by my increasingly irritated colleagues. Indeed, there are vendobiont fossils found in rocks thought to have formed in a deep ocean in Newfoundland. This was all the excuse I needed to visit that scenic Canadian Province. My wife and I had a wonderful visit staying in saltbox cottages and listening to Annie Proulx's quirky Newfoundland novel "The shipping news" as a talking book in our rental car. The principal localities of interest were Pigeon Cove and Mistaken Point, now within a UNESCO World Heritage Site, where coastal platforms expose hundreds of vendobiont fossils in life position. "You are calling this an animal, but is it justified to call it an animal? It looks rather plant-like", asks Sir David Attenborough in a 2010 television documentary at the Pigeon Cove locality. The same sentiments were echoed by every tourist group that first gazed in wonderment at the fossils of Mistaken Point as I was crawling over the outcrop to document their setting. The guide in the documentary, like those on the outcrop then explain that they could not be plants or even lichens because they lived too deep in the ocean for light to penetrate. The evidence they cite is numerous beds of sediment about a foot (30 cm) thick which change in grain size from gravel and sand at the base up to silt and clay at the top, superficially like deposits formed by deep submarine turbidity currents. This did not pass muster with me because I have seen numerous deep water turbidite beds in Oregon and New Zealand, and the Newfoundland examples were quite different in their red color, in having a sharp contact between the sandy and silty part, and in having very little clay. The rocks are beautifully exposed, but a better view was needed, so I sampled several complete beds and polished them like an ornamental stone to see the details of bed contacts and grading. With a complete sample also the beds could be analyzed for elemental composition, and the same geochemical depletions and volume changes in these Newfoundland beds as in South Australia are evidence that they also were soils. They are different kinds of soils than in South Australia, where the paleosols are studded with gypsum sand crystals and calcite nodules. The Newfoundland paleosols were non-calcareous and thin, with drab surface interfingering down into purple subsurface (Color Photo 12.2). Each paleosol was divided by a thin sandstone bed with sedimentary structures like those of tsunami beds familiar to me from the Oregon coast. The tsunami sands buried the fossils in life position, just as they buried the ghost forests of Oregon remaining from the tsunami of January 26, 1700. Additional evidence for coastal locations of the Newfoundland sections are a distinctive soil type with great masses of brassy pyrite nodules just below the surface. These kinds of soils are characteristic of mangrove and salt marsh habitats today. These intertidal paleosols have some

Color Photo 12.2 Ediacaran (565 million years) Maglona paleosol (Inceptisol) with drab top grading down to red subsurface at fossiliferous surface E in the Mistaken Point Formation, Mistaken Point, Newfoundland, Canada. The hole is from illegal collection of a fossil

of the simplest of Ediacaran fossils, the button-like *Aspidella*, which is similar in general appearance to modern maritime lichens such as *Verrucaria*.

The Newfoundland Ediacaran is like Oregon today in clear evidence not only of tsunamis, but of nearby volcanic activity. These volcanic tuffs provided the most compelling evidence that Ediacaran vendobionts lived on land. One of the Newfoundland tuffs had pea-sized accretionary lapilli and irregular scoria grains scattered within a silty tuffaceous matrix. This could not have been deposited in water, and especially not deep water, because the lapilli would have dissolved, and those surviving along with large scoria fragments would have sunk to the bottom of the bed. Other tuffs with scattered sanidine crystals up to a half inch (13 mm) long also could not have been deposited in water, because in that case the crystals would have settled to the bottom of the bed. Other tuffs again had tubular features from the escape of gases after they had settled, which is impossible for volcanic ash sifting down through the depths of the sea. In retrospect, this non-marine interpretation should not have been such a surprise, because a whole volcanic arc with rocks of the same age, about 565 million years old, has long been known in the next fault block to the west of the rocks with Ediacaran vendobionts in Newfoundland. This also is the age of a glacial advance in Morocco at a time when sea level was 600 m lower than before or after. Furthermore, the whole

Ediacaran succession in Newfoundland was deposited on continental granite, rather than on deep ocean basalt.

"But what about the evidence they could move from feeding chains and raking marks?" my increasingly irked colleagues asked. The rings and chains of impressions of forms like *Dickinsonia* and *Yorgia* have indeed been interpreted as successive feeding stations of a creature moving like a flatworm and dissolving away microbial mats by osmosis or cilial abrasion (remember they have no mouth or gut). Long streaks radiating from bulbous forms like *Kimberella* have been interpreted as raking traces by the feeding of a long proboscis. When I was able to see the relevant specimens some difficulties for these ideas were immediately apparent. The feeding chains are in lanes bumpy with microbial food, but the large near-circular areas between are smooth and barren of evident food. Should not the microbial food be less abundant in the feeding lanes than in intervening areas? The long marks emanating from *Kimberella* were not grooves with plowed up sides, but needle-like tubes extending outward. Can a proboscis extend needle-like outward for several body lengths? Both these striking fossils can be explained as periglacial soil features of the sort called frost boils and needle ice. Frost boils are areas where soil raised by freezing displaces lichens and microbial mats into narrow lanes of patterned ground between the water-flushed mounds. Ice needles form where a hollow plant has a water reservoir supercooled on the outside so that long needles extrude laterally into the soil or air. The more common kind of vertical palisade needle ice raising small rounded clods and larger divots of soil also are seen on the same slabs as the needle ice radiating from *Kimberella* fossils. These fossils did not necessarily move of their own accord, but may have been driven intermittently by wind on melting ice, like polsters of mosses and lichens in alpine and polar regions called "snow mice" or "vagrant lichens".

"What about *Dickinsonia* fossils with margins lifted from the slab as if about to become launched?" The curved divots missing from some *Dickinsonia* do not indicate they could rise in the floodwaters or seawater free of the bed, but rather that the forces lifting up the flap were stronger than internal body cohesion and less than forces attaching *Dickinsonia* to the bottom. Pieces could be peeled off leaving the rest attached, more like a sticker than a suction decal.

"How about the thin shims of sand covering *Dickinsonia* and other fossils but unknown for sandstones younger than Ediacaran? And will you never give up?" These shims, better called interflag sandstone laminae, are known now in many other sandstone sequences, and form by wind drift on fluvial levees, during exposure between flood events that produce the intervening thicker beds. The alternation of wind-blown and waterlain sediment is a characteristic of river deposits throughout Earth history.

"Surely isolation of cholesterol from *Dickinsonia* fossils means they had to be animals!" Cholesterol is indeed made by animals, but also by glomeromycotan fungi, which are already known from Ediacaran rocks as spores, vesicles and permineralized fragments. Not only cholesterol was found in the *Dickinsonia* fossils, but also the green algal biomarker stigmasterol. The bigger and presumably older *Dickinsonia* fossils had progressively more cholesterol and less stigmasterol. This surprisingly regular pattern would not be expected from an animal fouled with age by green algal overgrowth, or eating patches of green algae. The changing cholesterol/stigmasterol ratio is more like long term fungal growth using green algal symbionts.

Still more tests are being applied to this problem. The Ediacaran terrestrial lichen hypothesis has survived subsequent testing without a knockout punch so far. Many discoveries are so new that they are still filtering through the scientific community. Although some of my cocktail colleagues remain unpersuaded, consider some implications of this new view. Lichens are most common and diverse on land. They thrive on bare rocks and mountain tops unsuitable for most kinds of life other than microbial mats. Evidence for primitive microbial life on land extends well back into Precambrian time, as outlined in the next chapter. The Late Precambrian advent of lichens in a world of microbial earths would have been an important event for global change in ways that can be both observed and predicted. The biomass of lichen polsterlands is much greater than that of microbial earths and microbial rocklands. Compared with microbial mats, lichens also generate more acid and other compounds that promote weathering. Both increased biomass and weathering acid reduce carbon dioxide, and as a corollary of the greenhouse theory of climate, should have the effect of paleoclimatic cooling. The earliest large lichen-like fossils known are the "Twitya discs" of the Mackenzie Mountains in the Canadian Northwest Territories. Comparable discoid fossils have also been found in rocks of the same age in the Maly Karatau region of Kazakhstan. At about 800 million years old, these fossils are about the same geological age as a profound glaciation first reported by Sir Douglas Mawson in the Flinders Ranges, after return from his troubled Antarctic expedition of 1913. The bouldery glacial debris remaining from this glaciation has been found on every continent, and glaciers may have extended to unusually low latitudes at that time. Tropical glaciation is indicated by the way in which grains of magnetite and hematite lie parallel to the layers of sediment associated with the glacial deposits. At high latitudes these tiny mineral magnets stand up at a high angle to the surface of the Earth, but at low latitudes they lie flat. This is why compass needles need to be specially weighted in order to balance in Alaska or Antarctica. The implication of the flat-lying ancient magnetites of tropical lowland glaciers means that this was indeed a big chill. Joe Kirschvink of California Institute of Technology, who first documented this evidence of low latitude glaciation, has called this time Snowball Earth.

Other indications of late Precambrian global change come from large fluctuations in the isotopic ratios of carbon, strontium, and sulfur in sedimentary rocks dating back to this time. These subtle chemical traces also have been found to vary sympathetically worldwide. They indicate that several massive glaciations were accompanied by oxygenation of the atmosphere. Life on land may have played a role in abating an earlier Precambrian greenhouse by generating oxygen from photosynthesis in unprecedented volume. By sequestering carbon in biomass, consuming carbon as carbonic acid, and creating organic soil for erosion and burial, lichens may have revved up soil weathering as engines of global change.

These plausible effects of early lichens on land offer an even longer term perspective on the relationship between climate and life, again using the Proserpina Principle. On time scales of billions of years the Earth should have become progressively warmer. The sun grows brighter year by year as its conversion of hydrogen to helium grows greater in scope. The increase has been estimated at about 30% over the 4566 million years of Earth's existence. Each 1% increase in solar input should elevate Earth's temperature some 3–4 °F (2 °C), which adds up to 100–120 °F (50–60 °C) over the history of the planet. Despite these increasing inputs of solar radiation, Earth has become cooler through geological time. The primeval atmosphere of carbon dioxide warmed the Earth in the dim days when the sun was just beginning to fully ignite. But as the sun became scorchingly hot, carbon dioxide in the atmosphere has been consumed and buried by the activities of life, which include promotion of chemical weathering. Like grasslands and trees when they first evolved, early lichens would have been a carbon-hungry ecosystem that globally altered the balance of carbon and oxygen. Grasses and trees are curbed by cold climates, thus creating a negative feedback that prevents runaway cooling. It seems unlikely that lichens would have been severely affected by glaciation, because they thrive in glacial landscapes today. Their tolerance of cold and ice may be one reason for the severity of glaciation during the Late Precambrian, when glaciers extended to low latitudes. By this view, subsequent ice ages were less severe because the trees and grasses that induced these later ice ages were more vulnerable to freezing and glaciation.

Lichens remain common in glacial landscapes of the Arctic and Antarctic (Fig. 12.6, Color Photo 12.3). In the Dry Valleys of Victoria Land,

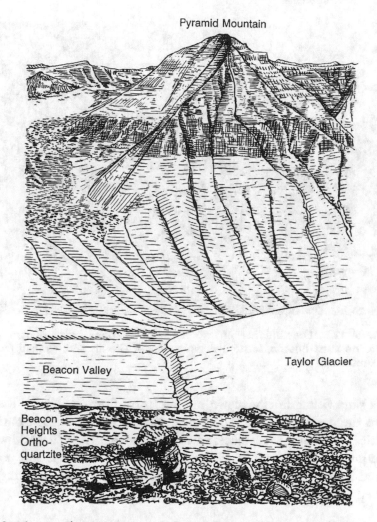

Pyramid Mountain

Beacon Valley

Taylor Glacier

Beacon Heights Ortho-quartzite

Fig. 12.6 View north across Beacon Valley to Pyramid Mountain and a lobe of the Taylor Glacier from West Beacon, Victoria Land, Antarctica. The boulders of quartzite in the foreground are colonized by endolithic lichens

Antarctica, lichens live within small crevices in sandstones. The red iron stain of the rock surface passes down to a thin (tenth of an inch or 2 mm) leached white zone, then a thin (tenth of an inch or 2 mm) green band of lichens before the gray interior of the rock. Summer temperatures in the Dry Valleys hover around −20 °C (−4 °F) and are seldom above freezing. In our summer month camping in the nearby Allan Hills, we had only a single day in which liquid water formed naturally outside. On that day, small drops formed at the end of icicles on sunny cliffs out of the wind. There are no large plants and few animals in this part of Antarctica. On one occasion a large seabird

Color Photo 12.3 Endolithic lichens (green and orange) on Devonian Beacon Quartzite on West Beacon, looking down to Taylor Glacier and Pyramid Peak, Dry Valleys, Antarctica

called a skua visited briefly, blown in by a storm from the nearby Ross Sea. Most of the time, the quiet and desolation were breathtaking. Continental Antarctica is one of the few places where one can experience a world without large animals and plants, but with a hidden biota of lichens and microbes comparable to that of Late Precambrian times. It was a beautiful living world even then.

LIFE'S ICY LEGACY

In vast Antarctic ice, I take some rest

In a yellow tent. A storm does its best,

Singing through guys and ruffling walls,

The liveliest thing in a world without calls

Of birds or bees. Yet this great ice desert

Does not lack life. Within rocks, covert,

Are lichens. Their rock-rind greenery

Is not apparent in ice and rock scenery.

And so in ages past, did life itself

Let oxygen cool air and glacial shelf.

Earth without cells and leaves of green

Would choke in sunlit clouds of steam.

13

When The Rust Set In

Early Earth like other planets such as Venus and Mars had very little free oxygen in the atmosphere. The timing and extent of transformation to the oxygenic atmosphere of today is revealed by color and chemical changes in paleosols.

Cape Wrath is an apt and evocative name for this storm-swept promontory in extreme northwest Scotland. It is a mountainous land of deep fiords and rocky tops flanked with purple-flowered heath and green soggy mats of *Sphagnum* moss. Felix Mendelssohn, the Jewish-German composer, captured the mood of these wild, wave-swept coasts well in his overture "Fingal's Cave," named for a sea cave not far away. He was captivated by this uncrowded land rich in Gaelic place names and ancient stones, a reminder of the Celtic cultural heritage of northern Europe. Across the mossy bog from the small village of Sheigra and down the rubbly slope of Cnoc an Staca in the sea cliffs and rock platform are exposed a pair of fossil soils. They are part of a hilly ancient landscape buried by red conglomerates and sandstones of ancient streams some 800 million years old. At that time, before the evolution of plants on land, this place would have looked even more barren and forbidding, but what was it really like? How did the Proserpina Principle mechanisms act in a world without trees, mosses or lichens?

We can turn to a variety of modern desolate landscapes for an idea of this ancient vista, but none exactly duplicate it. The barren volcanic landscapes of the North Island of New Zealand or of the Big Island of Hawaii are often used as models to depict Precambrian landscapes, complete with stormy seas and flashes of lightning around erupting volcanoes. There were undoubtedly

G. J. Retallack, *Soil Grown Tall*, https://doi.org/10.1007/978-3-030-88739-1_13

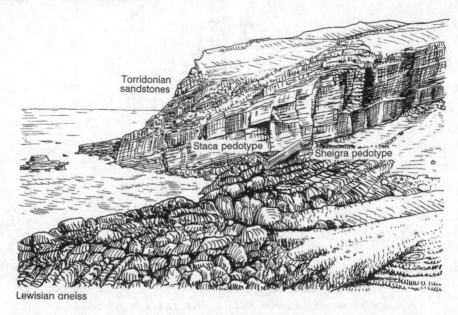

Fig. 13.1 Clayey paleosols of two different kinds (Sheigra and Staca pedotypes) some 800 million years old, buried by Torridonian quartzites of ancient streams, and formed on gneisses of a hilly landscape, on the north Atlantic coast, west of Sheigra, Scotland

local volcanoes, storms and lightning during the Precambrian, but the fossil soils of Cape Wrath are 10–12 ft. (3.0–3.7 m) thick and deeply weathered to clay. Nothing as exciting as a volcanic eruption or earthquake occurred here in hundreds of thousands of years of relentless and uneventful weathering.

Other landscapes used to visualize Precambrian environments are the vast sand plains of the channel country of outback southwestern Queensland, or the sandur plains of Iceland's coastal fringe. Precambrian rivers probably were broad and braided like these, but the thick oxidized, deeply weathered paleosols around Cape Wrath indicate that this was not a flat depositional landscape (Fig. 13.1, Color Photo 13.1). The paleosols include quartz-rich veins from the bedrock that are inclined at an angle to the top of the profile in the same way that trees and fence posts lean down slope due to soil creep (Fig. 13.2). This evidence for a hilly topography can also be seen by mapping this ancient land surface throughout the wider region around Cape Wrath. This was hilly landscape with 500 ft. (150 m) of relief formed footslopes to a mountain range that once existed to the west where the Atlantic Ocean is now. The ancient mountains are now buried in Greenland, and were torn away and drifted off with subsequent tectonic opening of the Atlantic Ocean.

Hot deserts of Atacama in Chile and Naseb in Namibia, and cold deserts of the Antarctic Dry Valleys also have been used to envision Precambrian

Color Photo 13.1 Neoproterozoic (800 million years) Staca (left) and Sheigra (right) paleosols (both Inceptisols) on amphibolite (left) and gneiss (right), covered by Applecross Formation near Sheigra, northwest Scotland

environments on land. But the shifting sands and silty texture of desert soils with their abundant nodules and crystals of calcite, gypsum and other salts are quite unlike the thick, clayey, non-calcareous to weakly calcareous paleosols around Cape Wrath. Deserts also have steep furrowed badlands and sharp wind-fretted rock peaks, but the ancient landscape was clayey and deeply weathered. Some 6 ft. (1.8 m) below the ancient surface are rounded remnants of fresh rock surrounded with altered rock like the corestones of a deep modern soil. Solid unweathered bedrock is as deep as 9 ft. (2.7 m) below the ancient surface. The abundant clay and carbonate nodules deep within the paleosol profile are similar to those of modern soils with a former mean annual rainfall of some 23–33 in. (550–850 mm), much more than in desert regions. It may have been even wetter, because the acidic leaching effect of modern soils comes in part from plants, whereas only microbes and rain were sources of acid in the Precambrian. The clayey paleosols are laced with thin veins of quartz-rich pegmatite which extend down into bedrock below. Pegmatite is rich in quartz and so less weathered to clay than the surrounding gneissic rock. Pegmatite slabs would have criss-crossed the clayey surface as a network of low walls and littered the rounded slopes. It would have been a stark and barren landscape, but eerily distinct from modern deserts in its

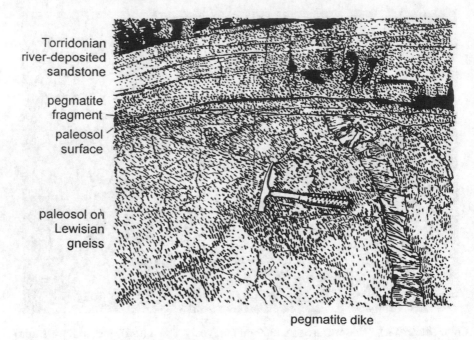

Torridonian
river-deposited
sandstone

pegmatite
fragment

paleosol
surface

paleosol on
Lewisian
gneiss

pegmatite dike

Fig. 13.2 The bending of quartz-rich veins into the top of the Sheigra paleosol is comparable to soil creep, and indicates that it formed on a well drained hillside sloping to the right

rounded hills and clayey surface. There are no landscapes exactly like this anymore.

These ancient soils do not fit comfortably in modern soil classifications. They lack the peaty surface of Histosols, the brown fertile surface of Mollisols, the clayey subsurface enrichment of Alfisols and Ultisols, the iron-stained quartz of Spodosols, the volcanic ash of Andisols, the relict crystal structure and bedding of Entisols, the ice-cracking features of Gelisols, the swelling clay structures of Vertisols, the profound chemical weathering of Oxisols and the shallow calcium-rich nodules of Aridisols. Only Inceptisols remain as an option, but the 800-million-year-old paleosols of Scotland are much thicker and more deeply weathered than modern Inceptisols. The Scottish paleosols can be accommodated in the already broad definition of Inceptisols, but not comfortably.

Here and there this Scottish landscape of 800 million years ago was probably tinged green from microbial mats similar to pond scum. Microfossils have not been found in these paleosols; nor would they be expected considering their degree of chemical oxidation and low concentrations of organic carbon. No microfossils are found in red and organic-lean paleosols of the past 400 million years either, even though these geologically younger pale-osols contain fossil root traces that would have relied on microbial support in weathering. Decay remains very effective still in well drained soils, like the Precambrian paleosols of Scotland. Good drainage is indicated for these ancient soils by their oxidation, by depth of weathering, and by deflection of the pegmatite veins at the surface of the paleosol in a manner comparable to soil creep in hill-slope soils. Nevertheless, the small remaining amounts of organic carbon in the paleosols show the same distribution as in modern soils, declining in abundance from the top to base of the profile. Furthermore, the carbon in this organic matter is isotopically light, a clear indication that it has been fractionated by photosynthetic organisms. Even at this remote time in Earth history these were living soils.

The other color seen everywhere would have been reddish brown. These ancient Scottish soils were pigmented with iron oxides and hydroxides, like many late Precambrian paleosols (Color Photo 13.1). Their weak red color and retention of iron indicates an atmosphere that contained common free oxygen. Elemental iron rusts to a red stain, and rust resists dissolution in water. We take this for granted now because as animals we depend on an atmosphere with 21% oxygen to breathe, and live among widespread red soils and rusting machinery. Similar red paleosols are known in even more ancient rocks (Color Photo 13.2). But it was not always so, as is clear from more ancient paleosols.

Paleosols of 2450 million years ago in the Elliott Lake area of Ontario, Canada, indicate a very different world than the strange Precambrian land-scapes of northwestern Scotland. Elliott Lake is a former mining town in the boreal woodland and rolling terrain of the Canadian Shield, now transi-tioning to a retirement community. Uranium was its main commodity, and its formation indicates unusual environments of the past. The uranium is found in placer deposits, which are local concentrations of the heavy dark mineral grains of uraninite in ancient river deposits. These rivers deposited thick sandstones of 2450 million years ago, covering paleosols developed on the ancient granites and greenstones. Gold also is found in the placers, drop-ping out of agitated river water along with the comparably heavy uraninite. Placer deposits of gold are known from almost every geological period, but not placers of uraninite. Uraninite is a very rare mineral at the Earth's surface today because it is easily oxidized in an atmosphere currently rich in oxygen. Uranium is soluble as an oxide, and once dissolved in water, it is transported and dispersed. Abundant uraninite grains in vigorously flowing streams are an indication that the atmosphere contained much less oxygen at that time.

Color Photo 13.2 Mesoproterozoic (1600 million years) green-red Daquan paleosols (Vertic Inceptisols) in the Bingmagou Formation near Wanghuazuang, Henan Province, China

Other indications of an oxygen-poor atmosphere come from ancient soils buried by the uraniferous Huronian sandstones. The buried soils of 2450 million years ago were clayey and extended down as much as 30 ft. into bedrock of granite and greenstone. The surface is the most clayey part of the profiles, with some mineral grains entirely altered to clay and thin skins of clay lining former cracks in the paleosol. Deeper levels of the paleosol profiles have rounded remnants of less clayey and less weathered rock, like corestones in deeply weathered modern soil. The clayey material around corestones in these profiles is an indication that water drained freely through deep cracks, carrying away dissolved products of weathering and filling cracks with clay in place of original feldspars and other minerals. Unlike modern well-drained soils, however, these paleosols have a gray-green color reminiscent of marsh and swamp soils in which oxygen has been excluded by stagnant water. How could this be, when clay skins and deep corestones are evidence that these Precambrian paleosols were not poorly drained marshes or swamps? They also lack the organic matter and distinctive minerals, such as siderite, found in swamp soils. For these paleosols, a lack of oxidation reflects a paucity of oxygen in the air some 2450 million years ago.

These Canadian Precambrian paleosols also defy classification. Inceptisol is again a reasonable option, but the depth of clayey alteration and very clayey

subsurface horizon of paleosol   surface of   overlying
with large corestones   paleosol   conglomerate

Fig. 13.3 A paleosol some 2450 million years old, overlain by conglomerate, tipped up on end, and then scraped clean in a glacial pavement from the last ice sheets of 15 thousand years ago, north of Elliott Lake, Ontario, Canada

surface is unknown in modern Inceptisols. Deep soil cracks and corestones are coated with iron oxides minerals in modern soils, yet these paleosols remained unoxidized (Fig. 13.3). This is so unusual that I consider them extinct kinds of soils, dinosaurs of soil classification. Green Clay is a suitable informal name for these ancient soils. I have also suggested the name Viridisol as a provisional extinct soil order, but hasten to point out that name is not approved as part of the U.S. soil taxonomy.

As different as they were from modern soils, they still were soils. Their abundant clay and gradational alteration down from the sharp upper contact indicate weathering by liquid water in a freely drained matrix at low temperatures and pressures. Under the microscope is an intricate internal pattern of shrinking and swelling, coating and leaching, and other soil forming processes. Also typical for soils is their chemical enrichment of aluminum and silicon and depletion of calcium, magnesium, sodium and potassium. Their isotopically light organic carbon is evidence of photosynthetic microbes. Iron sulfide seams like those made by bacteria are also found at the top of these paleosols. They really did form at the nexus of rock and air, and so are potential gauges of ancient atmospheres.

Using simple physicochemical models based on the chemical composition of a variety of ancient paleosols compared with the chemical composition of the granites and greenstones from which they formed, Dick Holland of Harvard University has estimated that oxygen was a trace gas in the atmosphere at that time. Although he left the effects of microbes out of his equations, the abundance of clay-like minerals and depletion of cationic nutrient elements and especially bioessential phosphorus in these paleosols is comparable with the results of biologically-mediated weathering by carbonic acid in modern soils. The source of the carbonic acid was probably respired carbon dioxide and atmospheric carbon dioxide available in amounts greater than today. At present, carbon dioxide is a trace gas, at levels of only 0.03% by volume, compared with 21% by volume oxygen. For the atmosphere of 2450 million years ago, Holland estimates that oxygen was no more abundant than carbon dioxide is today. There is also the argument outlined in the last chapter, that greenhouse gases are required to explain the profound weathering of these paleosols at a time of solar luminosity much lower than today. Such greenhouse gases could include methane and water vapor, as well as carbon dioxide, but the relative proportions of these gases and nitrogen in the atmosphere of 2450 million years ago will remain unclear until we know more about life in such ancient soils. What is clear is that oxygen was then a trace gas and carbon dioxide more abundant.

Landscapes of 2450 million years ago would have been eerie vistas of rolling clayey hills with stony litter. An important difference from the 800-million-year-old landscape of Scotland was the pale and bleached color of the soil, lacking red and brown hues. These Canadian hills also were probably dappled with green patches of microbes like pond scum, because of the isotopically light organic carbon and biogenic sulfide minerals detected in those paleosols. Life on land well back into geological time also can be inferred from the clayey and deeply weathered nature of paleosols as old as

3500 million years. Thick clayey profiles are commonplace in association with life, but difficult to create without life. Although life is commonly considered essential for soil, Stan Schumm of Colorado State University has proposed an interesting thought experiment concerning soil without life. In his hypothetical abiotic soil, each mineral grain loosened by weathering from bedrock should have been washed away by rain or blown away by wind shortly after it was free to so do. In this imaginary abiotic world, soils would be thin and sandy. But Precambrian soils are proving to be just the opposite: thick and clayey. Life is the glue that holds landscapes together for deep weathering. The filaments of simple unicellular cyanobacteria hold soil grains in place for weathering that releases needed nutrients. The slimes cast off by microbes for defense, protection and decay also bind the grains together so that weathering acids can continue their relentless attack. Even desert soils today are rich in bacterial crusts and scums, widely called cryptogamic earths or microbial earths. The grip of life on soils is ancient indeed.

Microbes are thought to have been involved in the great shift in atmospheric composition from a primeval greenhouse to the clear oxygen-rich skies of the later Precambrian. Many of the microbes found fossilized in cherts are similar to modern green pond scums that use photosynthesis to manufacture carbon compounds from carbon dioxide, releasing oxygen to air. They achieved much of this atmospheric revolution around 2400 million years ago which has been dubbed the Great Oxidation Event by Dick Holland, who based it on the changing oxidation state observed in many known Precambrian paleosols. The first to rust were granitic soils in which originally small amounts of iron were released by acidic weathering at rates slow enough for oxidation by small amounts of oxygen. Later to be reddened by iron oxidation were greenstone soils, in which so much iron was released by weathering that higher amounts of oxygen were needed to fix it as red oxides. In the sea also, the buildup of oxygen was retarded by iron-oxidizing bacteria to produce rusty red deposits called banded iron formations. These massive deposits of finely laminated iron minerals remain an important source of iron ore and date back some 1700–2500 million years. Once these sinks began to be saturated, oxygen accumulated in the atmosphere.

The thick paleosols on bedrock at Sheigra in Scotland and at Elliott Lake in Canada have carbon isotopic and organic geochemical evidence of life but little prospect to reveal the exact nature of life involved in the Great Oxidation Event. Sedimentary rocks and paleosols within sedimentary successions offer better prospects for preservation of microfossils. The Great Oxidation Event has traditionally been attributed to the evolution and spread of oxygenic photosynthesis by cyanobacteria common in shallow lakes and

Color Photo 13.3 Paleoproterozoic (2200 million years) Waterval Onder Vertisol from upper Hekpoort Basalt near Waterval Onder, Mpumulanga Province, South Africa

oceans of the time, as revealed by extensive laminated domes called stromatolites constructed by microbial mats of cyanobacteria and preserved in ancient limestones. But there are more complex fossils preserved in some paleosols of sedimentary sequences.

One of the best known fossiliferous sedimentary paleosols is the Waterval Onder clay paleosol some 2200 million years old in a deep road cut on national highway 4 just over the eastern scarp of the highveld on the way to Kruger National Park, South Africa (Color Photo 13.3). Andy Button of the University of Witwatersrand discovered this paleosol with its stunning deep cracks filled with sand from the river deposit above. It is still difficult to recognize Precambrian paleosols in sedimentary rocks because they lack root traces, and may not show marked soil differentiation. The Waterval Onder profile was so clearly a soil that it attracted the attention of many colleagues because it is so similar to a kind of soil still forming, a Vertisol or cracking clay soil. I have seen very similar soils in the cotton fields of Wee Waa, New South Wales, where I chipped cotton one summer as a teenager. The deepest cracks are on ridges of soil which outline swales, known by the aboriginal name gilgai (pond), because they fill with water when it rains. There is thus a cracking part of the soil and a ponded part, and it was in the ponded part

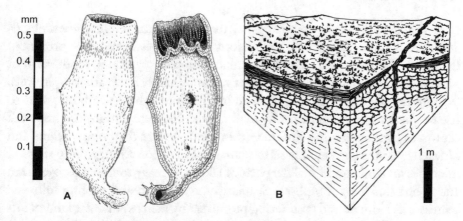

Fig. 13.4 Reconstruction of *Diskagma buttonii* (**A**), a likely glomeromycotan-nostocalean symbiosis, from the 2200 million year old Waterval Onder clay paleosol (**B**), a Vertisol of cracking clay from Waterval Onder, Mpumalanga Province, South Africa (from The Palaeobotanist, 2013, with Permission of the Birbal Sahni Institute of Lucknow, India)

that Andy Button noticed odd discoid structures. He and subsequent investigators floated a variety of ideas for these distinctive ellipsoidal forms, but their true nature did not become clear until I managed to get three dimensional images of them in the high intensity X-rays of the Lawrence-Berkeley synchrotron. They were more complex than vesicles, soil clods, sand crystals or nodules. The eye-catching ellipsoidal central hollow was defined by an organic wall that passed down into a stolon-like structure running along the surface of the soil. On top was a cup like structure with vertical filamentous structures (Fig. 13.4), but unfortunately metamorphic recrystallization obscured the biological nature of these filaments.

These complex little bladders may have a modern analog in the peculiar soil fungus *Geosiphon*, which is known from several modern German forest soils. *Geosiphon* is a small sausage shaped bladder with a large central ellipsoidal hollow and basal stolons and rhizines. The ellipsoidal hollow contains photosynthetic cyanobacteria such as *Nostoc* in a symbiotic relationship of producer and consumer. Some of my colleagues stop short of calling the cyanobacterium-fungus symbiosis of *Geosiphon* a lichen, because the cyanobacteria are engulfed within the cell (endosymbiosis), rather than imprisoned by surrounding haustorial hyphae as in other lichens (exosymbiosis). *Geosiphon* is a fungus of the Glomeromycota, a group of aseptate primitive fungi that includes mycorrhizae symbiotic with land plants, and responsible for greatly extending nutrient acquisition in soils. If *Diskagma* were similar biologically to *Geosiphon*, then its surface photosynthetic bladders and its subsurface network of stolons and rhizines would have promoted the Great Oxidation Event by both enhancing photosynthetic production of oxygen and increasing consumption of carbon dioxide by weathering in soils. Weathering consumes carbon dioxide in dissolved form as carbonic acid to release these nutrient elements by acid attack. Nutrient cations are lost with carbon as bicarbonate in solution within groundwater that joins streams and ultimately enters the sea. These fundamental fuels for a burgeoning mass of microbes on land and in shallow parts of lakes and seas promoted oxygenation and cooling of the atmosphere. Similarly, oxygenation and cooling followed increases in biomass and weathering promoted by lichens, vascular land plants and grasses. We humans were not the first organisms to change the world.

A big change in atmosphere like the Great Oxidation Event of 2400 million years ago required not only evolution of biological mechanisms for oxygenation, but big populations of oxygenators. Soils would have been important to that effort, because paleosols also are evidence for large areas of stable continents by 2600 million years ago (Fig. 13.5). Continents are mainly made of granitic rocks, coarse grained quartz-feldspar rocks, but looking back in time to 3800 million years ago granites are scarce. Before about 2500 million years ago the record is mainly basalt. A lot of it is pillow basalt formed when hot lavas are erupted into cold water, as has been observed many times in Hawaii. The earliest Earth represented by rocks was thus a global archipelago of small volcanic islands, which increased in complexity through time with addition of granites and other rocks. Early continental size can be evaluated from extent of granites, but paleosols provide another line of evidence. The first known calcitic desert paleosols (Aridisol) have been discovered in red beds 1400 million years old near Wolf Creek, Montana. A putative dolomitic paleosol 2600 million years old near Schagen, in Mpumalanga Province, South Africa has stable isotopic values more like marine rocks than paleosols. At the time of the Wolf Creek paleosol, large granitic continents had formed with marginal mountain ranges which kept the soil remote enough from moist maritime air masses to form calcareous desert soils. In contrast, geologically older soils were formed in humid maritime climates of small islands. As small continents became available for oxygen production and carbon dioxide consumption, weathering promoted by life effectively neutralized the steamy acidic primeval greenhouse to the clear blue skies that grace the Earth to this day.

Soil cyanobacteria and calcareous soil nodules are familiar in deserts today but what kind of life and weathering was there before 3000 million years ago in the anoxic atmospheres of the Archean era? Answers had been hiding in

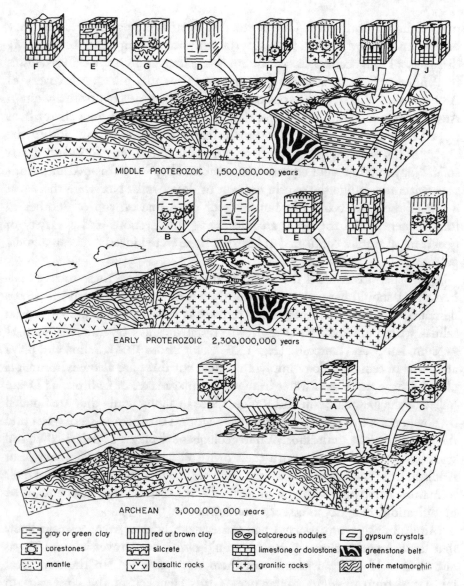

Fig. 13.5 A scenario for the evolution of soils through the Precambrian oxygenation of the atmosphere and growth of continents. A primeval volcanic archipelago in a steamy greenhouse atmosphere evolved into a planet with continents as life and soil consumed carbon dioxide and promoted cool and dry continental climates Soils illustrated are green clays (**A**), salty desert soils or salt-rich Aridisols (**B**), swelling clay soils or Vertisols (**C**), karst and drab cave earth or residual Entisols (**D**), oxidized incipient soils or Inceptisols (**E**), red deeply weathered soils or Oxisols (**F**), desert soils with silcretes or silica-rich Aridisols (**G**), and desert soils with calcareous hardpans or calcium-rich Aridisols (**H**) (reprinted from Retallack 2019, Soils of the Past, with permission of John Wiley and Sons)

plain sight for many years. One of the joys of studying paleosols is that they have often been overlooked by geologists without training in soil science. My discovery of Oligocene paleosols in South Dakota, and Ordovician paleosols in Pennsylvania were unforgettable moments for me, but the discovery of Archean alluvial acid sulfate paleosols in the Pilbara desert of northwestern Australia stands out as a red-letter day in 2007. They were grey cherty beds tipped on edge and fractured by many earth movements, but within the cherty layers were numerous sand crystals of the kind only found in desert soils. Not just a few sand crystals were visible. They were crowded into horizons within beds like the gypsic horizon of desert soils. Nor were there just a few of these paleosols, but hundreds of them one on top of another in long sequences. All the usual analyses using thin sections and a variety of geochemical studies confirmed at last that these were indeed Archean alluvial paleosols.

Later when I consulted the original published work on these rocks I realized that I should have suspected already. These old papers from the 1980s described Archean chert beds that formed as desert playa lakes or coastal salinas with numerous chert-replaced crystal outlines of minerals such as gypsum, barite and nahcolite (Fig. 13.6, Color Photo 13.4). Salina and playa are terms of sedimentology implying a lake, but these are also environments of soil. I have never seen water in the desert playa lakes of California's Death Valley or the Nevada Basin and Range. Once in a long while they are flooded by downpours, but mostly one can walk around on them with dry feet and the crunch of salts underfoot. While geologists see playas as desert lakes, soil scientists see them as soils with their own soil names such as Solonchak or Aridisol. Introduction of solutes and silt by rare floods is a minor part of their formation compared with the growth of sand crystals, cracking and other modifications that are, in essence, soil formation.

Mindful of the controversy I had started with Ediacaran paleosols, I knew that these Archean cherty paleosols would also be controversial, and it was not until 2016 that an account of them was published. In the same year my friend from previous conferences, Chris Heubeck of the University of Jena, also published with his student Sami Nabhan an account of very similar Archean paleosols from the Barberton region of South Africa. Chris also organized an excursion to these paleosols as part of the International Geological Congress in Cape Town in 2016, and I was able to see his marvelous discoveries and more of his incisive German logic for myself. This was thus another miracle year for me and a wonderful conclusion to a decade of work that had been mainly frustrating until then.

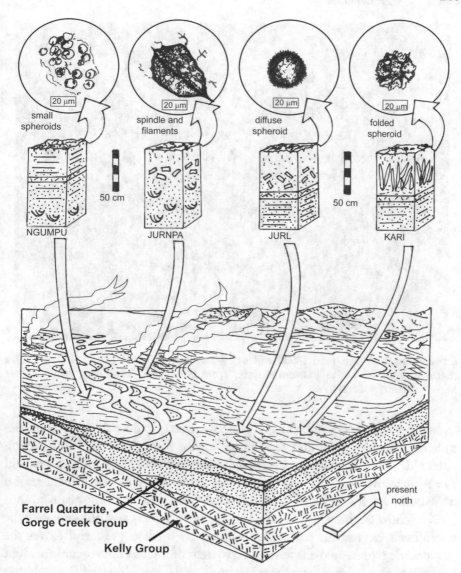

Fig. 13.6 A reconstruction of soils and microbes during deposition of the Farrel Quartzite 3000 million years ago in the Pilbara region of Western Australia. The permineralized soil microbes may have been actinobacteria (spindles, cf *Eopoikilofusa*), methanogens (small spheroids, *Archaeosphaeroides barbertonensis*) and photosynthetic sulfur bacteria (diffuse and folded spheroids, *Archaeosphaeroides pilbarensis*). (reprinted from Gondwana Geology, 2016, with permission from Elsevier)

Color Photo 13.4 Archean (3500 million years) Jurl paleosol (Aridisol) with silica pseudomorphs of barite (barium sulfate) from Panorama Formation, near Dresser Mine, Western Australia

Many rock units have these distinctive cherty paleosols with evaporite mineral pseudomorphs and they range in age from 3700 to 1800 million years old. The most informative though were from the 3000 million-year-old Farrel quartzite near the ghost town of Goldsworthy in the Pilbara region of Western Australia. These paleosols were at localities already made famous by Kenichiro Sugitani of Nagoya University in Japan, and showed that the microfossils he and his colleagues had documented so masterfully over the past decade represented a terrestrial ecosystem of microscopic organisms. Ken recognized five distinct forms of microfossil, and Chris House of Pennsylvania State University was able to supply carbon isotopic analyses of individual microfossils as clues to their likely metabolism. The small spheroidal forms (*Archaeosphaeroides pilbarensis*) had such isotopically light carbon that there is really only one likely metabolism for them as methanogens, those strange microbes of swamps and stagnant waters that produce methane or swamp gas. The large spindle-shaped microfossils (cf. *Eopoikilofusa*) are large hollow structures with small inclusions at the ends of tubular structures, and had very varied carbon isotope values, as if feeding on a variety of different organisms. This indication of respiratory metabolism and their distinctive morphology are matched well by actinobacteria, the most common soil decomposers besides fungi and animals. This confirmation that actinobacteria are fundamentally soil organisms, also answers another puzzle concerning other Precambrian paleosols. They must have had decomposers because they have only traces of organic carbon with photosynthetic isotopic compositions. Without decomposers they would have been black with carbonaceous materials like coals.

The larger spheroids (*Archaeosphaeroides pilbarensis*) have an unusual isotopic composition between that of methanogens and cyanobacteria, and very similar to that of photosynthetic sulfur bacteria. These bacteria are capable of oxidizing sulfur minerals to sulfate minerals such as barite and gypsum, and solve perhaps the greatest mystery of these paleosols. Why were there so many sulfate crystals requiring oxygen in paleosols formed under an anoxic early atmosphere? These sulfate crystals can be regarded as the residue of photosynthesis over the lifetime of the soil with creation of reduced organic matter by photosynthesis linked to oxidation of sulfur. In contrast, cyanobacteria create reduced organic matter by photosynthesis linked to oxidation of water to release free oxygen. Metabolisms have been destiny for our planet, because they control the chemical composition of water and air.

The Pilbara region of Western Australia also has been famous since 1983 for what were considered the oldest likely microfossils known by Bill Schopf the University of California Los Angeles, based on a remarkable assemblage of fossils from the 3460 million year old Apex chert near Marble Bar (Fig. 13.7). The fossils are in clasts within the chert, so redeposited from elsewhere. The chert is interbedded with basalts pillowed by eruption into sea water, so the fossils are probably marine. These interpretations and illustrations of the fossils can still be found in most textbooks, but were challenged in 2002 by Martin Brasier of Oxford University who had a very different view of both the fossils and their geological setting. He thought the fossils were not real, but artefacts created by displacement of abiotic organic matter by mineral crystal growth. Bill Schopf came back with a series of analyses indicating that they were not only organic, but independent of crystal boundaries. I have looked at these fossils in the transmission electron microscope and agree with Schopf that they trangress crystal boundaries so are not artefacts of organic matter pushed aside by crystallization. Brasier also considered the chert a hydrothermal conduit, but my geological mapping in the area supports the original interpretation of Schopf that the chert is an interbed to a thick sequence of marine pillow basalts. Brasier's conduit does not taper, and is not associated with any high temperature hydrothermal minerals or ores, and is

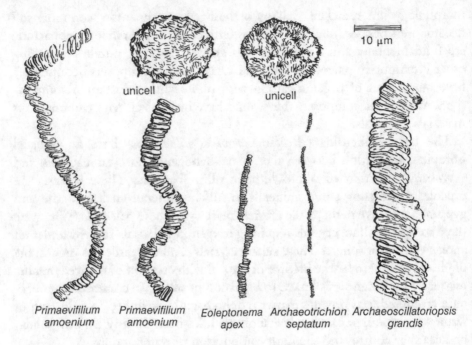

Primaevifilium
amoenium

Primaevifilium
amoenium

Eoleptonema
apex

Archaeotrichion
septatum

Archaeoscillatoriopsis
grandis

Fig. 13.7 Microfossils similar to photosynthetic bacteria some 3460 million years old from the Apex Chert of Western Australia. The scale is ten millionths of a meter

instead part of a widespread angular unconformity including a thick bedrock paleosol on the basalts.

To me the fascination of the Apex chert fossils is that they are indeed marine and yet so different from the terrestrial microfossil assemblage of the Farrel Quartzite and other cherty paleosols in the Pilbara region as old as 3458 million years. Microbial life in the ocean and on land was different from the very beginning of the fossil record. Spindles of actinobacteria characterize early anaerobic life on land, but cellular trichomes, perhaps of filamentous sulfur bacteria characterize the earliest known microfossils in the sea. Life on land and at sea may be equally ancient.

Even older evidence of life in the form of tufted mounds in limestones some 3700 million years old from West Greenland now look more like sedimentary laminae displaced by sulfate crystal growth. However acid sulfate paleosols in the same 3700-million year old rocks discovered by Nora Noffke of Old Dominion University have consistency of carbon isotopic isotopic composition suggestive of life on land (Color Photo 13.5). Some highly aluminous rocks in the same Greenland sequence as old as 3800 million years may be paleosols, but have not yet been fully examined. The further back one goes in the geological record, the more elusive are details.

Color Photo 13.5 Hadean (3700 million years) Isi paleosol (Aridisol) truncated upper left by quartzite and with black pseudomorphs of kieserite (magnesium sulfate), Isua Greenstone Belt, Isukasia, Greenland

Back beyond 3800 million years there is evidence from isolated crystals of zircon and it is even more tantalizing. Zircon is a robust mineral of zirconium silicate that also entraps uranium within its lattice. Analysis of the stage of decay of uranium has now been refined to such an extent that a geological age can be determined from a single crystal. Some of these zircon crystals date back to 4400 million years ago, and their isotopic compositions and mineral inclusions are evidence of a world of water and reasonable temperatures, less than boiling and more than freezing. These zircons formed in igneous rocks of that great age deep within the Earth and were eroded into the fluvial sandstones in which they are found in the Jack Hills of Western Australia. The Jack Hills sandstones are only 3000 million years old, and the really old zircons may have been buried and redeposited many times over. We no longer have their host rocks, or paleosols, or sediments to give context, and they formed somewhat removed from the surface of the Earth. Nevertheless, the Jack Hills zircons do indicate a remarkable continuity of surface conditions.

The Great Oxidation Event of 2400 million years ago was a turning point in the evolution of our planet, when its atmosphere diverged from the composition of atmospheres of every other planet and moon in our solar system. That transition was a biological engineering event from the ascendancy of oxygenic photosynthesis by cyanobacteria supplanting earlier anaerobic sulfur photosynthesis. We are not the first organisms to alter the atmosphere of our planet. Whether these organisms lived in soil or sea, they were fed ultimately by essential nutrients from weathering on land. Soil formation under moderate temperatures and with abundant free water has had a long and important role for our planet's atmosphere and life.

CHANGING COLORS

Soils are now red with iron-rich paint

Though stagnant water turns them faint,

Leaching their iron or keeping it blue.

Two billion years ago it was new

To see soils red or rusty stained.

Before then even soils well drained

Were entirely colored blue and green.

Such soils have not since been seen,

Drab and deeply weathered to clay

Down to rock corestones meters away.

Oxygen was so scarce in water and air

Soil iron was not rusted or kept there.

14

Soils in Space

Soils of other planets expand our view of the role of soils and life in planetary evolution. Mars, Venus, and our Moon provide an array of alternatives for reconstructing the evolution of weathering on the very earliest Earth.

My own taste in science fiction leans not so much toward the great achievements of intellect created by Isaac Asimov, Philip Dick, or Arthur C. Clarke, but to the pulp fiction comics of the mid-twentieth century with its strong visual imagery. This genre of popular art has been celebrated in the "Star Wars" movies. Like the imaginative artistic tradition of 1950s and 1960s pulp science fiction, most extraterrestrial creatures are recognizably like insects or humans, complete with legs, eyes, ears and mouths. The ageless stories of a son reconciling with his father, and of a commoner earning the love of a princess, owe more to Earth than the stars. The imagined landscapes of other planets are clearly inspired by barren deserts, polar ice caps and jungles here on Earth.

In contrast, landscapes now discovered on the Moon, Mars, and Venus are stranger than fiction. Soils on the Moon are produced principally by bombardment of tiny rocks. Soils of Venus form by processes more like pottery glazing than soil formation on Earth. These strange landscapes formed by sand blasting or by glazing have no clear analogs on Earth. They stretch the bounds of soil formation as it is commonly understood, and some have fairly asked whether they should be called soils. The US National Aeronautical and Space Administration (NASA) has had a simple and compelling reason for calling the surfaces of the Moon and Mars soils in press releases.

© The Author(s), under exclusive license to Springer Nature
Switzerland AG 2022
G. J. Retallack, *Soil Grown Tall*,
https://doi.org/10.1007/978-3-030-88739-1_14

They wished to be understood. Soils on both Mars and the Moon have both already supported human constructions, and perhaps eventually will also support plants and humans. Soils of other planetary bodies demonstrate that the most ancient known paleosols on Earth were not as strange as they could have been. They make us think again about soils and soil formation, and are the main subject of the new discipline of astropedology. They also reveal what soils could have been like on the early Earth. The rock record on Earth goes back only about 4000 million years. On the Moon and Mars, however, there are landscapes that ancient and rocks perhaps as old as 4500 million years. Some clayey and carbon-rich meteorites have been interpreted as products of ancient soil formation by aqueous solutions, and these are as old as 4566 million years, close to the age of the Solar System. Thus soils in space can reveal the very beginnings of soils here on Earth.

The first discovery of soils in space on July 20, 1969, transfixed my whole generation of baby-boomers born from prosperity after the second world war. I watched the moment that Neil Armstrong first set foot on the Moon one winter morning on television in the science classroom at my high school in Australia, at the same time as my future wife watched with her family late on a summer Illinois night. It was an historic moment that changed forever our view of the Moon. Subsequently we were treated to televised astronauts driving and bounding over the surface, and even hitting a golf ball off into the distance. Before that historic landing, there were real worries that the lander or the astronauts might disappear into quicksand, or raise a dust storm with every movement. Like soils of deserts, the lunar soil consists of particles ranging in size from silt to boulders. Almost all the "geology" of the moon has been inferred from study of large rocks in the soil. In all those Apollo missions that brought back moon rocks, not a single outcrop was sampled, only rocks in soil.

Loose soft soil covers most of the moon. The soil includes mineral grains such as feldspar. There also are small to large fragments of rocks made of several minerals. Such materials are found in soils on Earth as well, but lunar soils have very little water or air, and no clay or life. The minerals and rock fragments are essentially unweathered by water. These soils are thus more like the products of a road gravel quarry than a soil on Earth.

The really distinctive features of lunar soils are particles of glass and metal, which are rare in soils on Earth. The glass particles include sand-size spheres, dumbbells, and teardrops formed by the local melting of rock followed by cooling in the cold and extremely thin lunar air (Fig. 14.1). The impact of these tiny rocks and minerals melted the soil locally, sending up splashes of molten rock which then cooled to form the abundant glass particles that

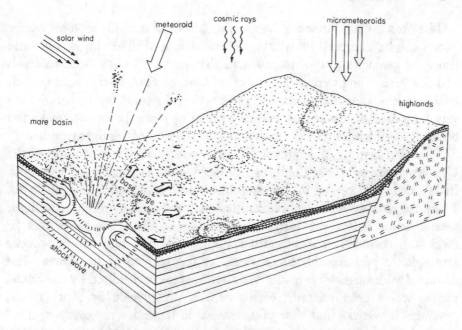

Fig. 14.1 Surface processes on the Moon (reprinted from Retallack 2019, Soils of the Past, with permission of the John Wiley and Sons)

litter lunar soils. Some of the glass particles look like miniature nut-coated donuts, because the molten glass incorporated other grains of the soil to form composite grains called agglutinates. The metal particles in the soil provide a clue to the origin of these various glassy particles in the lunar soil. The metal grains are like those of meteorites, and were added to the lunar soil by the constant rain of these tiny particles from space.

Soils on Earth are protected from a rain of small meteorites by an atmosphere that burns up incoming rocks smaller than a football. In the Moon's tenuous atmosphere, even microscopic particles hit the surface. They wreak havoc on spacecraft and astronauts as well, pitting and sand-blasting their surfaces. The tiny particles come down constantly like the beginning of a light rain; big rocks fall less frequently. Devastating impacts that created the large impact craters visible on the moon came at intervals of millions of years. Such huge impacts blast out expanding waves of boulders and soil that cover the soil with rubble. This sequence of events can be seen in the cores of soil obtained by drilling on the Apollo 15 mission. Separated by impact deposits in the 37 ft. (11 m) of recovered core are seven separate finer-grained layers rich in agglutinates and metal like the soil at the surface. These represent paleosols on the Moon, former lunar soils covered by successive impact deposits.

The time for soil formation by the steady rain of micrometeoroids on the Moon can be estimated both from the influx rate of these tiny missiles and from the geological age of paleosols in the Apollo 15 core. Both paleosols and the surface soil were hundreds of millions of years in the making. Left to themselves, those footprints and tire tracks will remain on the Moon for a very long time. I suspect they will be covered by more footprints, tracks and buildings well before they are obliterated by the rain of micrometeoroids.

That first landing of a human on the Moon was a great triumph of the U.S. space program, but the former Soviet Union also had some triumphs. In 1975, they succeeded in placing robotic landers on the surface of Venus. These rugged Soviet landers transmitted pictures and chemical analyses for several hours before the instruments melted at temperatures of 869 °F (465 °C), corroded in strong atmospheric acids, and were crushed under atmospheric pressures 96 times those at the surface of the Earth. The distorted wide-angle view of the Venusian surface transmitted by the Soviet probes was a great scientific event, even if not greeted by much media coverage. It was the first view of a new world beneath the heavily clouded atmosphere of Venus that had until then allowed some of the most imaginative flights of science fiction. It was not a planet of jungles and ancient civilizations, but a hazy barren land.

Both Venusian landers revealed what looked like sedimentary rocks, but were chemically like basalt. No great volcanic cones or lava flows could be seen. Curiously the deformation of the base of the spacecraft visible in the pictures indicated that these rocks were relatively low in density (1.2–2.5 g cm^{-3}) and porous (50–60%), more like soil than rock. Also unusual was the low electrical resistivity of the surface (only 73–89 Ω), and the darker color of the loose dust compared with the rock-like fragments. Dry rock or dust should have had high resistance to an electric current, and dust should have been lighter in color than rocks from which it was pulverized. Thus the simple rock and dust idea does not seem to work.

At one Venusian site, the rocks were rounded and smooth as if melted, whereas at another site they were angular and rippled. These differences can be interpreted as different stages of a bizarre soil-forming process that is more like the glazing in a potter's kiln than soil formation on Earth. The rippled angular slabs may be silt and other sediment washed downhill from the highlands by wind. Such a heavy atmosphere would have the power to transport larger particles than air on Earth. Between storms, dust and sand settled and was cemented and melted with salts like those of potter's glaze. Sulfate salts are likely because sulfur trioxide is the next most common component of the Venusian atmosphere after the ubiquitous carbon dioxide. Under the high pressures and temperatures at the Venusian surface, these salts would bind the loose sediments as electrically conductive rock-like masses. Over time the salts and some other minerals also would melt and ooze like glaze. How long this would take is unknown. If pottery kilns on Earth are a guide, it would require only hours or days.

This bizarre soil forming process has the effect of changing the chemical composition of rock surfaces. Such enrichment in sulfur is very different from depletion in sulfur and other life-nurturing elements on Earth. James Lovelock, British inventor, has argued that the lack of atmosphere on the Moon and carbon dioxide atmosphere of Venus is predictable from their size and the kinds of volcanoes observed on each, and thus indicates that both are lifeless. The strange soils of the Moon and Venus may be telling us the same thing.

The Moon and Venus have soil-forming processes about as different from Earth as can be imagined, but with Mars we are on more familiar ground. The images of the Martian landscape beamed back after landing of the U.S. Viking probes of 1976, Sojourner in 1997, Curiosity in 2012, and Perseverance in 2021 are similar to bouldery desert landscapes of Earth. With temperatures of −−130 to −22 °F (−90 to −30 °C) the Martian surface is even colder than Antarctic Dry Valleys. Soils in barren landscapes of Antarctica are thought to be similar in some ways to the soils of Mars. We will not be entirely sure until missions return samples of Martian soil or until we send humans to examine them firsthand. Our information on Martian soils comes entirely from the pictures, experiments, and analyses provided by robotic landers.

For many years mainly surface views of Martian soils were available, but all that changed when Curiosity rover beamed back images of a foot thickness (30 cm) of soil profiles from a locality dubbed Yellowknife Bay in Gale Crater (Fig. 14.2, Color Photo 14.1). Beneath a rock table mantled with sandstone are clayey deposits with many characteristic features of desert paleosols considered some 3700 million years old from crater counting in the surrounding region. Like paleosols on Earth these Martian profiles have a sharp top above a cracked massive layer, then a layer with abundant nodules of hydrated calcium sulfate (bassanite), and below that a silty layered deposit. Two such sequences reveal soil formation covered by layered sediment, which was then subjected to additional soil formation. These profiles with abundant sulfate are comparable with playa lake soils in extreme deserts like Death Valley, California. Especially characteristic are sulfate filled hollows and cracks in the rock. The rounded hollows find a match in vesicular structure characteristic of modern desert soils. The cracks taper downward, are filled with sulfate, and form an interconnected network, defining a network of clods that

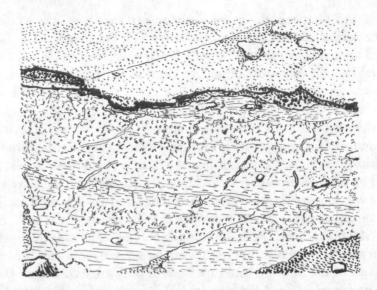

Fig. 14.2 Paleosol profiles 3700 million years old in Gale Crater, Mars

Color Photo 14.1 Noachian-Hesperian (3700 million years) Yila paleosol (Aridisol) with common nodules of bassanite (calcium sulfate) in Sheepbed member, Yellowknife formation, Yellowknife Bay, Gale Crater, Mars (courtesy of NASA)

becomes looser upwards. The vertical cracks also are wiggly, as if compacted by overburden of a lot of rock removed by erosion, and unlike other forms of geological fracturing.

Technological improvements in the Curiosity rover's instruments also delivered chemical and mineral data on these profiles with accuracy comparable with those from a laboratory on Earth, and a vast improvement over the Viking and Opportunity instruments. These data make clear that there was significant chemical and mineral alteration by weathering which converted most olivine into smectite clay. The accumulation of hydrated calcium sulfate salts (bassanite) in nodules and vesicles is another product of water-based weathering. Salt and clay are natural breakdown products created by weathering reactions in water on Earth. Carbon dioxide dissolved in rainwater creates carbonic acid, which then attacks primary minerals such as feldspar, converting them to clay and salts. Stronger sulfuric acid is often involved when soils on Earth are littered with sulfate salts. It is difficult to envisage how these processes could be occurring in the extreme cold today on Mars, where liquid water is not stable. In addition, the Martian atmosphere is very thin, with less than 1 per cent of the Earth's atmospheric pressure. Although carbon dioxide is 95% of that thin atmosphere, there is enough to create carbonic acid if it were not too cold to rain.

Nevertheless, there is evidence that Mars did not always have such cold, thin air. Martian orbiters have now documented both the topography and surficial geology of much of the planet. Landforms that at first looked like the pyramids of Egypt and Elvis Presley's face have been found to be ordinary looking hills in high-resolution images now available. The age of different landscapes can be estimated from high-resolution airphotos by the density of impact craters: the older the surface, the more impact craters. Some landscapes 3000–3500 million years ago old show enormous braided stream channels like those eastern Washington State, where silty sediments were scoured back to expose basaltic bedrock like giant scabs. The Channeled Scablands of Washington formed by enormous floods of water released some 12,000 years ago from the failure of an ice dam impounding a large glacial lake backed up past Missoula, Montana. Similarly on Mars, outwash channels reveal a time of catastrophic flooding, but 3000–3500 million years ago. Even more ancient and heavily cratered Martian terranes some 3500–4000 million years old show narrow sinuous valleys like those of Earth formed by collapse of rock weakened by groundwater sapping. There was free water during this earlier phase of Martian prehistory. The paleosols of Gale Crater formed during this wetter time, some 3700 million years ago, and then were preserved in a deep freeze. Many exposed surfaces of Mars are paleosols out of equilibrium with the current frigid climate.

If life existed on Mars it would have been back then, rather than now. Possible evidence for life on Mars has recently been found in the form of microfossil-like shapes in meteorites from Antarctica. A handful of the hundreds of known meteorites collected on Earth are thought to be fragments of the Moon and Mars because they differ from other meteorites in their geological age, magnetic characteristics, and chemical composition. The meteorites thought to be from the Moon are identical to rocks brought back by the U.S. Apollo missions. Some of the really big impacts on the Moon and Mars blasted rocky debris into space and into an Earth-crossing orbit. Microscopic structures, minerals, and organic compounds suggestive of life have been found mineralized in the cracks of one of the meteorites with a Martian chemical composition. Few rocks have been subjected to such intense scientific scrutiny as a supposed Martian meteorite found in the ice fields around the Allan Hills in Antarctica. Studies of radioactive minerals in the meteorite show that the crack-filling minerals have a geological age of 3600 million years, but the parent rock crystallized 4500 million years ago, and includes minerals shocked by impact some 4000 years ago. The life-like microscopic structures are small cylinders, both isolated and in chains like those of bacteria. Also found were tiny discs and spheres of magnetite, like those produced by some bacteria. Chemical compounds such as polycyclic aromatic hydrocarbons in the meteorite are also like those produced by life. Each of these lines of evidence is problematic because the Antarctic ice fields are not as completely devoid of life as commonly supposed, and because, despite impressive precautions, laboratory contamination of the meteorite remains possible. The microscopic cylinders are much smaller than any life form known on Earth, so small that it is difficult to conceive of all the molecular machinery of a functioning cell fitting within such a small space. Nevertheless, our search for better evidence of life on Mars has been invigorated by these discoveries. The conclusive demonstration of life beyond Earth will be a momentous scientific discovery.

Also preserved in the deep freeze of space are relics of an even earlier phase in the evolution of soils in the solar system, again from meteorites, which are space rocks fallen to Earth in the fiery passage of a shooting star. Meteorites can be found, though rarely, on the ground, but for the freshest samples it is best to follow the trajectory of shooting stars. Many meteorites are distinct from rocks of Earth in having small spherical mineral grains called chondrules. These are called chondritic meteorites, and their spherical grains were produced by melting at high temperature and pressure. A few meteorites have chondrules scattered in a matrix of carbon compounds and iron-rich clay, with occasional veins of salts. This variety of meteorites is linked by a curious fact: dating by decay of their radioactive isotopes shows that almost all of them are about the same age, some 4566 million years old. Only the handful of rock-like meteorites thought to be from Mars and the Moon have younger geological ages. There are no rocks as old as 4566 million years on Earth, because they have been destroyed by weathering or by mountain building. Rocks approaching this age are rare on the Moon and Mars. This early date is thought to be about the time that the solar system became differentiated from a great whirling cloud of dust. Meteorites are a clue to this early phase when precursors of the planets were differentiated from the primeval cloud that eventually became the sun and planets.

The main source of meteorites is the asteroid belt between Mars and Jupiter. There may have once been a planet there, shattered by a cataclysmic collision in space, but it seems more likely that aggregation into a large planet was prevented by the enormous gravitational attraction of nearby Jupiter. The origin of meteorites in small planets or asteroids accords well with their varied composition. By this view, the metallic ones came from the core of large planetismals, and rocky meteorites were forged in the high pressures of large planetismal interiors. Most meteorites with chondrules however have neither the interlocking crystalline texture nor abundant metal of large planets, and were probably from asteroids no larger than 300 miles in diameter. Asteroids today have surfaces very similar in physical and optical properties to chondritic meteorites, which may be remains of the most ancient soils in the solar system. The Dawn Mission of NASA arriving at Ceres in late 2015 has provided copious images and other data on an icy planetismal with surface clayey and carbonaceous soil, and subsurface salts, like carbonaceous chondrites (Color Photo 14.2).

An intriguing puzzle of carbonaceous chondrites is their mix of salts and clays formed at low temperatures, on the one hand, and on the other hand, chondrules of minerals formed at rock-melting temperatures of 2200–2500 °F (1200 1400 °C). The chondrules are thus a parent material, and were formed by electrical discharges or collisions in space. Salts and clays on the other hand, formed by water-hosted weathering of the high temperature minerals to low temperature soil minerals. Thin slices of carbonaceous chondrites viewed under the microscope show many details comparable to those found in soils. Some of the chondrules are clear and fresh, but many are embayed or rimmed with iron oxides and clays. Some are mere ghost-like outlines in the fine-grained matrix. Such progressive weathering is often observed in soils. The veins of salts, calcite and dolomite in carbonaceous

Color Photo 14.2 Primeval (4566 million years) surface soils of carbonaceous chondrites with white spots after salts in Occator Crater on the asteroid Ceres (courtesy of NASA)

chondrites cut across chondrules with their weathering rinds and also across the nearby matrix. These are like cracks within a soil filled with crystals from evaporating water. The clays locally form small curved layers, as if they were filling cavities in a soil in successive episodes of infiltration. Other clays are randomized in orientation with local streaks of preferred orientation like those produced by shrinking after drying, then swelling with wetting of soil. The high content of carbon compounds, including such complex molecules as amino acids and sugars, is also evidence for processes involving water at low temperatures and pressures. Rounded clumps and microscopic threads of concentrated carbon in carbonaceous chondrites have been interpreted as evidence for extraterrestrial life. As for putative Martian microfossils, most scientists remain skeptical of this interpretation. Nevertheless, carbonaceous chondrites, Martian paleosols and Antarctic Dry Valley soils are rich in salts and swelling clays, and may represent a soil type widespread during formation of the planets and asteroids of the solar system.

These various observations and comparisons confirm that some carbonaceous chondrites were ancient soils rather than cold condensates of dispersed gas, liquid and dust from space. All of these distinctive alteration features would not be created in meteorites passing a little too close to the sun. Nor are these features formed by soil formation after they fell to Earth. Like all meteorites, the carbonaceous chondrites have a rind of melted rock from when they burned and crashed to Earth. The rind is hard and heat resistant,

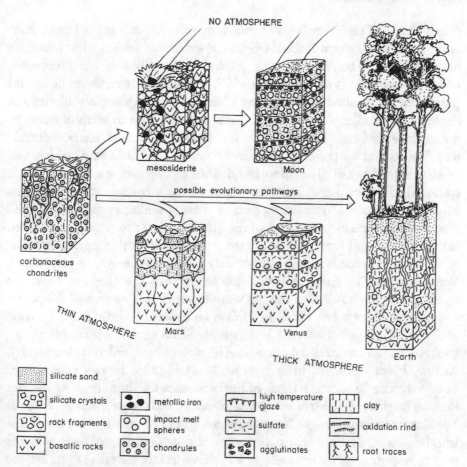

Fig. 14.3 Schematic comparison of soil formation on Earth, Moon, Mars, Venus and some hypothetical meteorite parent bodies (reprinted from Retallack 2019, Soils of the Past, with permission of the John Wiley and Sons)

protecting the meteorite from weathering and overheating. Carbonaceous chondrites are so fragile and soil-like that they can be crushed between the fingers. They do not last long in the open. Most of the specimens studied have been found soon after their fall or were frozen in large ice fields. They are not products of weathering during or after fall to Earth, but fragments of soils from the very beginning of the solar system.

Our view of soil is profoundly changed if the primeval soil was a carbonaceous chondrite rather than a rocky barren dust like that of the Moon (Fig. 14.3). Clays, salts, and organic matter are full of potential, fertile and complex. They are plausible planetary management systems neutralizing carbonic acid from water and carbon dioxide released by volcanic and other vents. On small planetary bodies formed by collision such as the Moon, gases and liquids seeping out of the interior were exhausted within a few hundreds of millions of years. The force of gravity on these bodies was insufficient to keep gases near as an atmosphere. Their soils then became sterilized and pulverized by bombardment of tiny meteoroids. On planetary bodies not depleted of volatiles, such as Ceres, mass was sufficient to retain a transient atmosphere, and soils like carbonaceous chondrites formed. Carbon dioxide was flushed out of their atmosphere in rain as carbonic acid, which was neutralized by partial dissolution of minerals to clay and salts. Atmospheric carbon dioxide, methane and water vapor was also consumed by formation of carbon compounds. Consumption of these greenhouse gases cooled the planets against the increased heat of the sun. On Mars this process allowed an early period of soil formation in a watery planet, but with progressive atmospheric loss to space, much of the planetary surface froze over in permafrost. Venus is more like Earth in size, much larger than the Moon or Mars. Its thick atmosphere so close to the sun would have deeply weathered the planet early in its history, but life either failed to evolve there or failed to cope with a growing greenhouse crisis. Earth is within a zone where weathering could be maintained throughout a long history of increased solar luminosity. At first, Earth was saved from greenhouse overheating by neutralization of carbon dioxide in soils like those of the carbonaceous chondrites. As the sun became brighter still, Earth was saved from the hellish greenhouse of Venus by microorganisms hungry for greenhouse gases. As photosynthetic organisms became successful in their efforts to draw down carbon dioxide, they fell prey to respiring microorganisms which converted them again to greenhouse gases. Mass extinctions from asteroid impact and the spread of glacial ice also prevented life from depleting greenhouse gases entirely and tipping the Earth into the permanent snowball that is Mars. Soils, then life, kept Earth in habitable balance. It has not been a perfect balance, as the succession of ice ages and greenhouse paleoclimates demonstrate, but the dynamic thermostat of the Proserpina Principle allowed escape from the lifeless terminal icehouse or greenhouse of our planetary neighbors.

SKINS OF PLANETS

Moon, Mars, Venus and asteroids

Have softened surfaces. Can we avoid

Calling these altered rocks a soil?

Yet so different is the constant toil

Of meteoroids pecking at the Moon.

The oven-like air of Venus will soon
Melt dust and rock into shining glaze.
Old soils of Mars have salts and clays.
But more like soils of earliest Earth
Are carbon-rich chondrites. Life's birth
Consumed them here, but their places
Are frozen asteroid's dark faces.

15

Living Soil

The origin of life has been envisaged within small ponds of ocean water and deep-sea hydrothermal vents, but a case can be made for evolution of life within, and as soil.

The idea that life came from soil is arguably the most ancient human view of the origin of life. A cuneiform text dating to 2000 BC from the Sumerian city of Nippur in present-day Iraq, describes a feast of the gods, hosted by Ninmah (mother earth) and Enki (god of the waters). Exuberant from much food and wine, Ninmah created six different kinds of humans out of clay, while Enki decreed their fate and gave them bread to eat. An ancient Egyptian bas-relief dating to about 1400 BC in the birth room of the temple of Luxor depicts a human body and soul fashioned out of clay by Khnum, a ram-headed god of the Nile River waters (Fig. 15.1). Such beliefs are understandable in lands such as Egypt and Iraq where frogs and other creatures emerge from the soil after floods. Similar views are found also in the original (Yahwist) portion of the Judaic biblical account of *Genesis*, dating from 900 to 500 BC. The name Adam means "red clay" in Hebrew. The name Eve is the Hebrew verb "to be". The Genesis account thus implies that Eve came from Adam as "living soil."

The origin of life from soil entered the stream of scientific thought through the ancient Ionian Greeks. Xenophanes (510–475 BC) wrote that "all things that come to be and grow are earth and water." According to Diogenes (ca. 440 BC), Archelaus thought that "living beings were generated from the earth when it is heated and it throws off slime consisting of milk to serve as a

© The Author(s), under exclusive license to Springer Nature
Switzerland AG 2022
G. J. Retallack, *Soil Grown Tall*,
https://doi.org/10.1007/978-3-030-88739-1_15

Fig. 15.1 Ancient Egyptian god of the Nile, Khnum ("Lord of the Dark Waters"), modeling the future king Amenhotep III, as both ba ("personality") and ka ("vital essence"), on a potters table, from a bas-relief in the Birth Room at the Temple of Luxor, Egypt

source of nourishment." The idea of the spontaneous generation of life from soil remained popular for centuries, endorsed by such influential thinkers as Aristotle (ca. 384–322 BC) and Lucretius (ca. 99–55 BC). But the whole idea was abandoned in the late nineteenth century when Louis Pasteur showed that life did not arise spontaneously from the soil or organic matter. Pasteur demonstrated experimentally that soil and organic matter could be sterilized by heating. New life emerged from soil and organic matter only when viable propagules could be seen under the microscope. Since then the idea that life originated in soil has become a minority view, but one to which I will return after consideration of competing theories for the origin of life.

Especially appealing is the idea that "life like Aphrodite was born on the sea foam" in J. D. Bernal's memorable line. Charles Darwin popularized life's origins in a "warm little seaside pond". The idea that life evolved in the sea can be traced back to the Ionian philosophers Thales (ca. 585 BC) and Anaximander (ca. 565 BC). The Aegean Sea still teems with life, much more obvious and easy to observe than life in the soil. An aquatic origin of life received a scientific boost from a stunning experiment in the early 1950s. As geologist Harold Urey was preparing to depart his laboratory at the University of Chicago for a scientific conference, he arranged for his graduate student Stanley Miller to begin their planned experimental simulation of the Precambrian world. Miller took a system of sealed flasks containing water gently heated by burners to simulate the sun. The gases in this completely enclosed microcosm included carbon dioxide and ammonia, but no oxygen because the atmosphere of the early Earth was thought to be poor in oxygen. The whole mix was sparked with electricity to simulate lightning. Amazingly, water in the bottom of the flasks accumulated a dark, tarry mixture of organic molecules, including amino acids and sugars. These kinds of experiments have been tried many times since with a variety of gases. As long as oxygen is scarce, the result is more or less carbon compounds. The experiments have also been run with clay and metals in the solution, as would be found in tidal flats, sea bottoms, and sea cliffs. Yields of carbon compounds are promoted by these substances, so that the relevance of the experiments to the origin of life in the sea, as opposed to soil, are now less clear. These and other experiments have never resulted in a freshly created microbe emerging within a sterilized test tube, but are impressive confirmation that some of the basic building blocks of life could be assembled under likely surface conditions on the ancient Earth. These experiments confirm how organic-rich sludges could have formed, like the fine-grained matrix of carbonaceous chondrites, which dates back to the earliest stages of the origin of the solar system.

Discovery of volcanic vents during exploration of the deep sea in recent decades has suggested yet another set of theories for the origin of life. In the dark world of the deep ocean, where the great basaltic underpinnings of the sea floor are pulling apart, there are chains of volcanic chasms that erupt lavas and streams of seawater heated by contact with the molten rock. Hot water billows out of these vents like black smoke because of the rapid precipitation of sulfides and other dissolved substances into cold sea water. The vent fluids are like toxic waste, highly acidic, oxygen starved, and extremely hot (temperatures up to 716 °F or 380 °C). There is little organic carbon or organisms immediately around them, but inches away the sea floor is crowded with life. There are large white clams and crabs, and peculiar pale tube worms. At this great depth there is no possibility of photosynthesis, so the community is fed by microbes capable of feeding from chemicals precipitated from rapidly cooling vent waters. On early deep sea expeditions by Oregon State University, Jack Corliss was very impressed with these bizarre hydrothermal vents, and he proposed that spongy rock and vent deposits around the toxic stream were a likely site for the origin of life.

Another possibility is that life evolved elsewhere in the universe and colonized our planet as propagules that could withstand long-distance transport in space. This intriguing concept of "panspermia" goes back to the turn of the century and the Swedish chemist Svanté Arrhenius. A related idea favored by science fiction writers and scientists such as Francis Crick is deliberate colonization of the Earth by advanced extraterrestrial civilizations. All manner of organisms could have been sent, ranging from influenza viruses, to unicellular bacteria, to sophisticated aliens in space vehicles. Such views have some appeal in this age of space exploration, but they are not especially useful. They merely remove the question of the origin of life to another planet. The environment where that life evolved is likely to have been Earth-like in many respects because life has long been well suited to our planet. Thus it remains useful to consider the origin of life from natural causes here on Earth.

If the extraterrestrial origin of life can be put aside, then how to choose between these separate concepts of an origin of life in soup, spa, or soil? The French molecular biologist Jacques Monod suggested a simple and elegant criterion. How well do each of these proposed sites for the origin of life explain a fundamental paradox of life itself, summarized in the title of his slim book "Chance and Necessity"? The problem is that even the simplest life forms are incredibly complex. Yet they need to be complex in order to function. The origin of such complexity by chance is so unlikely as to be nearly impossible, but at the same time it is absolutely necessary. The enterobacterium *Escherichia coli* is one of the simplest known organisms, yet still quite complex (Fig. 15.2). It has only one cell and no differentiated nucleus. The DNA that codes for replication of this organism is strung along nucleoids on a tangled skein of internal partitions. The DNA itself is a complex molecule with thousands of atoms arranged in a sequence that specifies the structure of the cell. Instructions of DNA are read, activated and deactivated by proteins, also of complex structure. The probability of even such a simple organism arising by random movement of atoms is more remote than one chance in a number close to the number of all the electrons in the visible universe: an incredible long shot.

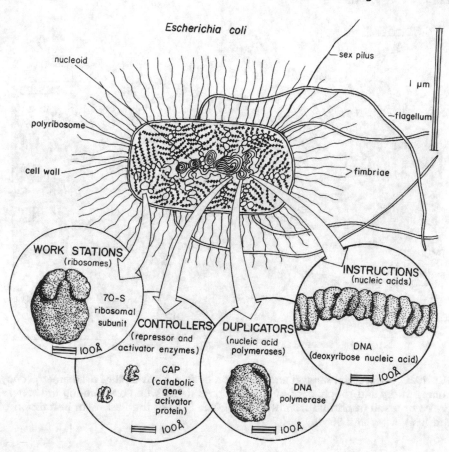

Fig. 15.2 The gut bacterium *Escherichia coli*, a simple organism, yet a complex machine of organic molecules. Complexity is necessary for it to live and work, but defies the odds for creation by chance, unless natural selection played a role (reprinted from Retallack 2019, Soils of the Past, with permission of the John Wiley and Sons)

Natural selection is a biological process that defies such long odds. This was Charles Darwin's seminal insight in his landmark book "On the Origin of Species," published in 1869. The odds of evolution of an eye or a brain are also very long. In a world where the blind or stupid are eaten, eyes and brains are a matter of necessity rather than chance. What was it about pre-life and the earliest life that was selected and converted chance to necessity? We may never know the exact answer to this question, but a good case can be made for a single adaptation fundamental to soil—stickiness. Clay, organic matter and life are glues that bind the soils, and protect them from erosion. In an imaginary world where non-clayey and non-carbonaceous soils are eroded by wind and water, soils would quickly become clayey, carbonaceous and finally

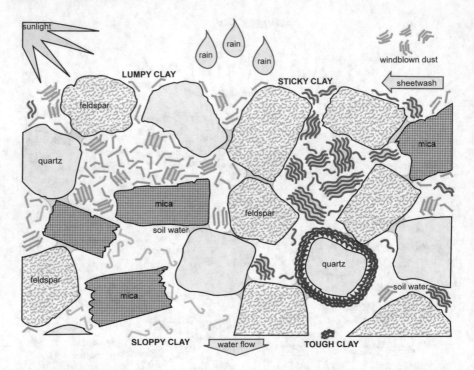

Fig. 15.3 Differential survival and evolution of four kinds of clay, nicknamed Sloppy, Lumpy, Sticky and Tough, each of different consistency when wet, among the mineral grains of a soil (reprinted from Retallack 2019, Soils of the Past, with permission of the John Wiley and Sons)

living. The Scottish chemist, Graham Cairns-Smith, has proposed several stories along these lines that reveal even more explicitly how the early Earth would select for soil and life (Fig. 15.3).

Imagine an early Earth in which temperatures and pressures were moderate enough for liquid water. Carbon dioxide choked the atmosphere but its greenhouse warming effect was then useful in offsetting the chill of an early sun, whose thermonuclear fires were not yet fully ignited. As the sun became brighter and harsher, warm rains dissolved carbon dioxide to make a dilute solution of carbonic acid. This acid would be neutralized by dilution within large bodies of water, but more slowly neutralized by reaction in soils, which soak up enormous quantities of water. This acid weathers soil rocks and minerals, dissolving salts to be carried off in groundwater and leaving clays as a weathering residue. The warmer it became, the more energetically carbonic acid weathered the rocks, consuming carbon dioxide in the process. Less carbon dioxide was utilized if temperatures fell so that water froze to ice and was unable to continue the work of weathering. An early balance was achieved between atmosphere and rocks that resulted in soils that were a mix of clay, minerals and water acting like a crude chemostat and thermostat. Such neutralization of a primeval acidic atmosphere may have kept Earth's surface temperature within the narrow bounds of liquid water.

In such an environment we can move to stage two, when clays formed by weathering were not just any clays, but rather those clays best suited to the amount of rain and warmth, and the nature of the minerals being weathered. Following Cairns-Smith, we can avoid the daunting complexity of clay chemistry, by calling them Sloppy, Lumpy, Sticky and Tough. These clays were all weathered from mineral grains of a well-drained soil, intermittently wet by rain. Sloppy clay was like swelling clay in Vertisol soils. It expanded dramatically when wet, and contracted when dry, but was so rich in expanding layers that it fell into small pieces after rain and was carried down deep into the soil. Tough clay on the other hand was virtually inert like the clays in Oxisols. As the mineral grains were weathered to this clay, it formed a tough and immobile rind that eventually sealed off the grain from further weathering. Between these extremes were Lumpy and Sticky. Both clays expanded when wet, thus bridging the crevices between the grains so that they were not carried away by the wind and percolating rain. When the soil dried, they cracked away from their parent grains exposing more fresh mineral surface to the weathering solutions that created more clay. Lumpy clay formed small clumps of clay and protected the soil from erosion in a mild rain. A clay that was less easily dislodged and more uniformly coherent, such as Sticky, was needed to defend the soil against thunderstorms and sheet wash. This simple weathering system had a strong component of natural selection. Lumpy and Sticky quickly dominated, because Sloppy and Tough were either washed deep into the profile or eroded by wind and water. The future belonged to Lumpy and Sticky, which had physical properties that protected the soil against erosion so that it persisted at the surface in the zone of materials and energy transfer where more clay was made. Under such a positive feedback system, soils bound for posterity were soils bound by clay. They remained uneroded and became more clayey day by day.

In a third stage of early soil evolution, there were widespread clayey soils producing a variety of clays. In our mineral melodrama of Lumpy and Sticky, Lumpy prevailed in the long term, because of Lumpy's ability to form small lumps that rolled and dispersed for short distances over the soil. Few of the clays became extinct, because of differences in local rainfall and drainage. Nevertheless, the ability to form propagules, no matter how crude, was an advantage. An intriguing thought of Cairns-Smith, as yet without scientific confirmation, is that clays could be what he calls a "genetic crystal." Particular sequences of layers with different chemical composition or particular patterns of extraneous atoms within the clay lattice could conceivably determine with some predictability a particular degree of lumpiness, stickiness, or expansion when wet. Such compositionally determined physical attributes can be considered adaptations to particular environments. Such concepts only now are beginning to be tested with ultramicroscopic investigations. Regardless of how these researches pan out, the parallels between living creatures and a working soil are clear. Both soil and life are complex interfaces, maintaining a dynamic equilibrium that is self-sustaining, taking and giving back materials to their environment.

This theoretical "clay life" can be envisaged as a starter system, still far from life as we know it based on organic matter, which characterizes stage four. Organic matter forms naturally in watery environments like those of the early Earth in which oxygen is scarce, as shown by Stanley Miller's experimental synthesis of organic matter, but we need an unusual concentration of organic matter to make an organism. It is here that the ideas of life's origin on the seashore or in deep-sea hydrothermal vents fail utterly. The world has so much water that early oceans would have been a very thin soup indeed. Charles Darwin's envisaged a pond that was small, and the smaller the better, with optimal size range down to the size of small crevices in soils. Seashores are also impermanent, shifting environments, subject to erosion or to sedimentary cover by storms or tsunamis. Deep sea vents also are prone to collapse, or become starved of hot fluids by unpredictable shifts of the Earth's crust. Organic matter on land however would have been sticky. Like molasses on a cornfield, any kind of tarry organic matter would have saved the soil from erosion. The more organic matter, the longer soil persisted to form more organic matter. Such stability becomes self-reinforcing. With roving wind and water always plucking at the least organic soil to erode and carry away to the sea, it would not have been long before soils of the early Earth were both clayey and carbonaceous. But simple organic compounds like those of the Urey-Miller experiment are not the basis of life, large molecular weight polymers such as sugars and proteins are needed, and these need wet-dry cycles and dehydration to form.

Long-lived carbonaceous soils could have gained organic matter even by cumbersome methods of Urey-Miller synthesis, because over extended time spans of soil formation even inefficient methods of synthesis are effective. One well known, but inefficient way of forming organic compounds from carbon dioxide is to excite that gas with light to transfer oxygen and electrons to nearby reduced iron in minerals. Thus photo-oxidative processes creates opaque iron oxide minerals as a shield from ultraviolet and other radiation that would otherwise break down complex organic molecules. As discussed previously, we have a fossil record of such early soils, full of clay, opaque oxides and organic matter. They are the carbonaceous chondrites, known from isotopic dating to be some 4566 million years old, as old as Earth itself.

Complex organic life is still a long way from this primitive tarry clay. Stage five sees the beginnings of organic complexity in the nooks and crannies of the soil. Clay has a smooth appearance, but under the great magnifications possible under the electron microscope, it is a jungle in there. Some clays at a microscopic level are like fluffy corn flakes arranged into houses of cards. Other highly magnified clays look like ragged wisps and tufts of grass. Others again are like long worm-shaped accordions of flakes. Opaque iron oxides also have a fantastic array of ultramicroscopic shapes: tubes, balls and flakes. Graham Cairns-Smith compares these structures to a chemistry laboratory, full of flasks, retorts and tubing. The synthesis of organic molecules requires a variety of different steps: add a little alkali, then perhaps a little acid, dry out, filter, irrigate, and so on. Clayey soils have the compartments and wet-dry cycles that promote formation of really complex organic molecules, melded together from smaller organic molecules. Not only iron but molybdenum and other metals mainly available in soils are critical to many biologically significant macromolecules. In contrast, the idea that life originated in the sea or submarine volcanic vents fails because the ocean is a dilute solution of relatively uniform chemical composition, with all irregularities evened out by mixing toward a lifeless chemical equilibrium. Soils on the other hand seldom reach chemical equilibrium. Ever more complex long chains of organic slime are more and more effective at binding soil, warding off erosion and promoting further synthesis of clayey soil and organic matter.

By stage six, clay lost control to organic compounds, and an important corner was turned towards its current role as a matrix and provider of life. By this time long chains of organic matter began to form structures of their own. The balls, sheets and strings of organic matter provided alternative chemical laboratories to the natural enclosures provided by clays and iron oxides. The reactions are driven by energy from the warmth of the sun and trickling of rain into cracks of the soil, but guided by organic catalysts, templates, and containers for organic synthesis. These intermediary organic materials would have been comparable to what molecular biologists call ribosomes in organisms. Concentrations of organic matter would have dissolved and dispersed in the extended medium of the sea and in hot volcanic vents. Such organic balls, strings and sheets also have been found in carbonaceous chondrites, and so were present early in the evolution of terrestrial planets.

Stage seven was the appearance of organically-based genetic molecules. Each soil is different in the grains available for weathering, and the amount of rain and warmth. Too much organic slime in the soil would dry up and roll away in flakes. Too little would leave the grains vulnerable to wind and water erosion. There would have been strong selection for the right kind and amount of organic matter for each location. Those soils lacking organic matter did not last. Long complex molecules such as RNA and DNA may initially have played a role in binding the soil. The advantage of such molecules over other kinds of organic matter is their ability to replicate accurately by splitting down their ladder of bases, and then gathering the complementary base sequence. What counts in the long term is consistency, and this was the advantage of the molecules of heredity. Neither DNA nor RNA were likely to have thrived in the sea. Some of the components of these molecules require synthesis under acidic conditions, others under alkaline conditions. All are rapidly degraded at the high temperatures found in volcanic vents. Each component of these complex molecules is joined by a hydrogen bond that requires dehydration.

Stage eight sees another important step toward life. Organic molecules coded not only for themselves, but for other organic molecules as well. The consistency of replication in their own chemical composition and physical characteristics was now transferred to organic strings, tubes, balls and sheets in the soil. This was the beginning of a revolution, a genetic revolution. The outmoded crystal machinery of the soil was increasingly outpaced and outclassed by the speed and efficiency with which organic molecules could commandeer the primary source of energy from the Sun and the primary materials of carbon dioxide, water and nutrient elements. Organic matter had now learned to live in that excited skin of the Earth we call soil.

By stage nine, the society of genes and proteins had divorced itself from its clayey matrix by encapsulation within walls of cells. This is perhaps the most difficult step of all. Even with potential genetic molecules and organic materials all around, the organization of random molecules into a functioning cell was a triumph of natural selection against long odds. Cells bind and cells move. They glue the soil better than molasses and tar because they are more responsive to changing conditions, both within the soil and beyond. Some of these independently functioning cells may have continued in the ancient mode of producing organic matter by a variety of chemical alterations of carbon dioxide using energy from the sun. Others may have begun consuming the reserves of organic matter in the soil. At first they respired only a small amount of the soil's organic matter because too greedy a feeding frenzy led to soil destabilization and erosion. Eventually, however, the cover of microbial earth was so effective and quick that both production of organic matter and its decay were done by living cells. The old carbonaceous chondrite soils began to be cleansed of organic matter formed in sluggish and archaic ways. The revolution was complete and cellular life had earth firmly in its grip. Weathered clays were no longer so important to organic synthesis, but were open grocery shelves to be pillaged of nutrients. The splash of rain, the warmth of sun, and nutrients of minerals were now husbanded mainly by living cells.

This stage in the evolution of life was achieved at least as long ago as 3800 million years. There is evidence in the isotopic composition of organic matter of that great age for a preference for the lighter isotope of carbon (^{12}C rather than ^{13}C) characteristic of photosynthetic life and its isotope-fractionating enzyme rubisco. There are no older rocks of a kind suitable to preserve these isotopic ratios, so life may be older still. The oldest little-metamorphosed rocks, some 3500 million years old, include plausible permineralized cells similar to bacteria and other simple microbes, as described in the previous chapter. These particular microfossils have been disputed as artefacts of crystal growth and bubbling in a volcanic hot spring, but their carbon appears organic in isotopic composition and in other ways. They also show tubular decay patterns similar to that of modern actinobacterial decay. Other evidence for microbial decomposers comes from paleosols 3500 million years old of two kinds. Some on bedrock were thick, clayey, and relatively clean of organic matter content, as well drained soils and paleosols of have been ever since. Without decomposers, soils would have been dark and carbon-rich if supporting any life at all. Others are full of sulfur salts and relatively carbonaceous with microfossils, including both producers and consumers.

There remains a huge gap between carbonaceous chondrite as plausible soils of 4566 million years ago, and the familiar isotopic compositions, fossil microbes, and paleosols 3800–3500 million years ago. The scenario outlined above to fill in that gap would seem highly unlikely were it not that each step is improved soil stability that serves to perpetuate the system within the zone of energy and materials transfer. The evolution of microbial consortia included evolution of even more effective soil binders. There was the engulfing of the originally free-living precursors of nuclei, chloroplasts, and mitochondria that resulted in bigger microbes with complex internal organelles. The competition for a place in the sun was heightened again by the societies of producers and consumers that we call lichens. Then came flat blades of non-vascular plants, the deep and ramifying roots of trees, and the thick sod of grasslands. Dinosaurs, cities, and nations thrived as long as soil remained productive. By this view, the complexity of life did not evolve by chance, but by natural selection for mechanisms to hold a place in the soil within the zone of surficial materials and energy exchange.

LIFE'S ORIGINS

The paradox of life is its complexity,

Too much for chance, but not necessity

Of natural selection. What was chosen

Before genetic codes were frozen

As basic to the branching tree of life?

A simple starter system was its wife

And mother. No other than soil itself

Could spark a prelife and give it health.

Carbon-rich slime and clay in the ground

Gives immunity from erosion all around.

When life came it played the same role.

To sun, mineral, and water, it kept hold.

16

The Proserpina Principle

Plants cool the planet by consuming carbon dioxide, whereas animals warm by breathing carbon dioxide. This Proserpina Principle extended back to microbial earths, and prebiotic weathering, which were also important Gaian regulators of Earth's environmental stability.

This book has been a journey around the world and deep into geological time to test ideas about life and soil in past cycles of global climate change. One guiding idea outlined in Chap. 1 is what I have called the Proserpina Principle: plants cool the planet, but animals warm it. Carbon dioxide I exhale is fixed as carbon by plants of my garden, thus reducing atmospheric heat capacity. When I eat and digest vegetables, I breathe out their carbon to thicken the atmosphere with carbon dioxide, thus increasing heat capacity of the air. The breathing and growth of organisms is important on time scales of years and decades, but burial of carbon in soil and sediment is needed to regulate atmospheric levels of carbon dioxide over millennia. Coevolution of plants with animal consumption, and of animals with plant defenses, extends well back into geological time. The evolution of thick woody tree trunks some 370 million years ago cooled the planet, because wood is difficult to digest, and much of its carbon is buried in soils and sediments. The evolution of termites some 230 million years ago warmed the planet, because, by digesting wood and mining the soil, they reduced the amount of woody fiber buried in soils and sediments. By this view, we are not the first organisms to warm the climate of our small blue planet. Nor were termites the first to change our

G. J. Retallack, *Soil Grown Tall*,
https://doi.org/10.1007/978-3-030-88739-1_16

world. Meteorite impacts and large volcanic eruptions also have brought on fearfully destructive global warmings.

Soil has a special role in cycles of life and climate, beyond that of mere matrix for animals and plants. Acidic solutions of carbon dioxide from soil animals, mostly microscopic in size, liberate from mineral grains the primary nutrients, such as phosphorus, that fuel plant growth. Thus a host of microscopic creatures also play a role in our atmosphere, and it is technically better to speak of consumers warming the air with respiration and producers cooling the air with photosynthesis. There also are contributors to global climate change beyond soils and life. Plants are aided in cooling the planet by absorption of carbon dioxide by marine plankton. Animals are aided in their warming effects by carbon dioxide and other greenhouse gases gushing from volcanic fumaroles. Large volcanic eruptions and meteorite impacts also destroy the local carbon management of soils and plants, creating atmospheric greenhouse crises. These crises of destructive, elemental forces of nature are shortened by soils that work to regulate our water and air with a global democracy of all beings, a fundamental balance of nature.

The balance of producers and consumers, like the composition of the atmosphere, is not static, but a dynamic equilibrium of cycles of life and global change. Atmospheric cycles have been particularly well revealed by global monitoring of atmospheric carbon dioxide over the past 50 years. This has demonstrated a general rise from 280 to 418 parts per million of carbon dioxide with the burning of fossil fuels. In addition to this upward trend, which has tracked global warming and fossil fuel burning, atmospheric carbon dioxide analyses also reveal annual variations in concentration of this greenhouse gas of 5 parts per million. In the northern hemisphere, carbon dioxide rises from its lowest value each fall as trees shed their leaves in a blaze of color. The rise in carbon dioxide proceeds with winter respiration of animals and microbes in the soils, as well as from us and other animals on the soil. Carbon dioxide levels reach a peak during the spring, as trees once again bud into leaves that begin a long summer of photosynthesis and atmospheric carbon dioxide reduction. An important clue that this annual fluctuation in carbon dioxide is biologically controlled rather than purely celestial, is less marked annual variation in the southern hemisphere (Fig. 16.1). If simple differences of solar radiation with Earth's rotation were the sole explanation, southern hemisphere fluctuations in carbon dioxide should be the same as in the northern hemisphere. In contrast, southern carbon dioxide fluctuations are more muted because there is less land, soils are poor, and deciduous plants are rare in the southern hemisphere. The Australian outback has dry, phosphorus-poor soils with evergreen eucalypts and acacias, due to a

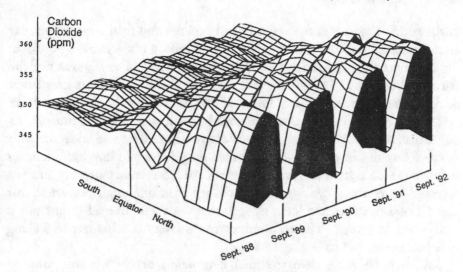

Fig. 16.1 The Proserpina Principle of annual fluctuation in carbon dioxide of the atmosphere is revealed by this graph of carbon dioxide abundance at various latitudes for four successive years. The effect is most marked in the northern hemisphere where soils are fertile and many plants deciduous. In the southern hemisphere, the fluctuations are in phase with southern seasons, but subdued because so many plants there are evergreen and soils are poor in phosphorus and water (Reprinted with permission from Volk 1997, Gaia's breath, with permission from Columbia University Press)

long geological history of weathering and erosion. Australian soils were not renewed by mountain building and glaciation that was widespread in the northern hemisphere. Southern hemisphere carbon dioxide levels are high in spring and low in fall, and so synchronized with southern seasons, but out of phase with northern seasons. Northern hemisphere carbon dioxide fluctuations follow northern rhythms of life and soil production, but the southern hemisphere follows southern rhythms. Large differences in severity of harsh northern and mild southern seasons are not simply consequences of tilt of the Earth's axis and its orientation to the Sun, but rather of soil productivity and geographic differences between the hemispheres. As a young Australian mountaineer, I had experienced life-threatening blizzards in the Australian Alps, but never the deadly chill of my first winter scurrying between buildings during blizzards on the northern Illinois till plain. These harsh temperature fluctuations in the north are plainly reflected by fluctuations in atmospheric carbon dioxide, which are more exaggerated in the northern than in the southern hemisphere.

We began with the great annual cycle of planting and harvesting celebrated in ancient times as the return of Proserpina from her winter sojourn in the underworld. Then small mysteries of soil nodules and firm clays caught our attention as clues to the role of soils and paleosols in understanding cycles of life on longer time scales. Later we considered the role of soils in the rise and fall of civilizations and of human ancestors, and in the evolution of grasslands, of dinosaurs, of forests, and of life itself. We delved into books, drove miles and miles, marshalled the might of sophisticated scientific instruments, set our thrusters to the moon and Mars. We have travelled far and wide, not only in space but also in time. We have drilled deep holes and dug long trenches in baking tropical heat, and frigid Antarctic cold. Soils and paleosols turn out to reveal more about cycles of global change than one might expect at first sight. There was much to learn right beneath our feet. Marcel Proust put it well, when he wrote, "The real journey of discovery consists not in seeking new landscapes, but in having new eyes."

Soil is much more than the final destination of every heroic journey. The unprepossessing rind of soil is the wellspring of our daily sustenance and of ourselves. The soil is deceptively placid and passive, yet always busy, breathing, rotting, and churning. The soil barely resists the mark of the plow, the shaft of the water well, or the gaping holes of quarries, but soil is relentless in patching and repairing these wounds. The fury of the storm, the surging of tides and the great crises of nations do little to alter the cycles of decay and renewal that create soil. In a world of emergencies and deadlines, soil continues to build clay and humus. Soil is the real estate of nations and of civilizations, which prosper by its bounty, or die scattered within it. Dinosaurs and millipedes have come and gone by grace of the soil. Soils play an important role in maintaining the surface environment of the planet we call home. Soils may have been where life evolved, the mother of all. We are in a sense a highly evolved form of soil, doing what soil does, but more quickly and efficiently.

Soils are home not only to us, but to algae and amoebae, lichens and millipedes, trees and termites. Most of these cooperatives of production, consumption, recycling, filtering and building are small and dispersed, but some are impressive for their size and organization. Consider for example the reddish mounds of fungus-cultivating termites that dot the plains of Africa and Australia. These small pale insects work in vast coordinated colonies to create a system of underground galleries, which give them protected access to water and to twigs and other organic foodstuffs. A complex array of earthen chambers within the mound is used for cultivating fungal food from organic matter, and for rearing young from the eggs of a single bloated queen. The coordinated action of millions of termites in a single colony can be regarded as a superorganism. Winged workers beat a breeze through the mound for air conditioning. Increased air flow and the fungal gardens create chemical conditions within the mound different from surrounding soil, allowing precipitation of calcite in otherwise calcium-poor soils. Calcite-encrusted termite nests are the only local source of agricultural lime in regionally infertile soils of many African countries. Not only do termites alter their home environment, but their insatiable demand for plant material cleans the soil of organic matter. Some tropical soils have been through the guts of termites many times over, and almost completely cleaned of organic matter in the process. The respiration of termites is a significant source of global carbon dioxide. It has been that way since termites and other wood-eating insects evolved millions of years ago, when dinosaurs ruled the Earth.

The idea of Earth itself as a kind of superorganism like a termite nest that regulates its surface temperatures and other conditions to be favorable to life has been called the Gaia hypothesis by James Lovelock. Gaia was mother earth of Greek mythology. As the firstborn from the primeval darkness of Chaos or Erebus, she was mother of Ouranos the sky and Pontus the sea. From her union with Ouranos came the first generation of gods, the Titans. One of these was father time or Kronos, who sired Zeus. He and the other Olympian gods rose to power only after a protracted struggle. During these struggles that reflect ancient Greek religious and political crises, Gaia intervened by providing weapons and shelter for her favorites and by creating monsters to combat her foes. Gaia thus predates the Olympian goddesses of wifely duty, of passionate love, of career women, of home and hearth, of grain, of maidenhood, and of wild nature. Hera, Aphrodite, Athena, Hestia, Demeter, Persephone, and Artemis were all part of the Olympian pantheon ruled by Zeus. Gaia represents something more fundamental and archaic.

Gaia has been widely identified with the idea of a great goddess of Neolithic to Paleolithic times, before the invention of writing. Evidence of an ancient, omnipotent grand matriarch has been sought among archeological records, especially among female figurines (Fig. 16.2). Among my favorites are the obese torsos with stylized heads of Paleolithic age, some 30–40,000 years old. The so-called "Venus of Willendorf," named for its Austrian discovery site near Vienna, but without any known link with the Roman goddess of love, presents an unflattering yet compelling image of woman as a roving incubator. More closely allied to the Greco-Roman Aphrodite-Venus is the Canaanite goddess Astarte or Ashtoreth, the pagan goddess reviled by authors of the Bible's Old Testament. She is often shown with domestic animals and plants, the bounty of harvest. Yet another potential earth mother is found among pottery figurines some 4300 years old in the ruins of the early Bronze Age town of Mohenjo-daro, Pakistan. These urbane figurines

Fig. 16.2 Some early images of goddesses: left, Late Palaeolithic "Venus" of Willendorf; center left, mistress of domestic animals and plants from the lid of a late Bronze age cylindrical box from Ugarit, coastal Syria; right center, early Bronze age mother goddess from Mohenjo-daro, Pakistan; right, late Bronze Age snake goddess from Knossos, Crete

have loin cloths, necklaces and elaborate head-dress, which may have been designed to hold incense sticks or offerings. Snake-toting goddesses of late Bronze Age Crete, some 3650 years old, are often interpreted as a prototype for the Greco-Roman warrior goddess Athena-Minerva. Like the Mohenjo-daro figurines, these also have been found in domestic shrines. The snake has been interpreted as a symbol of power and wisdom, perhaps because of the hallucinogenic trances and death brought on by its venom, or perhaps because of the role of snakes in guarding grain stores from rodents. Birth, home and death are indeed fundamental and ancient themes.

Marija Gimbutas has argued that these are all manifestations of the same archaic great goddess, who for want of a better name could be called Gaia. It is easy and comforting to imagine a time before the biological role of men in reproduction was understood, when it seemed that women alone ruled over birth, life and death, and such a religion still has undeniable appeal. However, these various representations could equally be minor deities, portraits of real people, initiation figurines, good luck charms, puppets, dolls, priestesses, witches, birthing aids, worry stones, or boundary markers. Like all great art, the statues elicit our hopes and dreams, yet remain aloof and silent. Classical authors such as Herodotus (ca. 450 BC), Diodorus (ca. 36 BC) and Apuleius (ca. 125 AD) mention matriarchal societies and goddess cults in Egypt, Crete and Anatolia, but by then they were not widespread. Sumerian mythology is the oldest actually written down, rather than preserved by oral tradition. Although Sumerian goddesses have a prominent role, they are not completely the center of attention.

Fig. 16.3 Earthrise from the Moon. The contrast between a barren Moon and watery Earth full of life suggested to James Lovelock the Gaia hypothesis

The Gaia hypothesis for James Lovelock was something quite different from a mythic earth mother. For him it is a scientific hypothesis, or set of hypotheses, that Earth's surface environment is regulated by life processes. Lovelock is an independent British inventor and atmospheric chemist. His Gaia hypothesis arose from fascination with the very different planetary atmospheres studied during space exploration of the 1970s. Of the many images returned from exploration of the Moon and nearby planets, perhaps the most evocative was a blue, cloud-draped Earth rising over the barren rocky horizon of the Moon (Fig. 16.3). Other planetary bodies such as Mars and Venus have silty to bouldery soils lacking water, and air full of carbon dioxide. Lovelock pointed out that if Earth can be imagined without life and at chemical equilibrium with gases released by volcanoes, then it should have an atmosphere mainly of carbon dioxide (99%) and minor oxygen (1%) and an ocean like a paste of sodium chloride (35 vol %) and sodium nitrate (1.7%). In contrast with this expectation, the atmosphere of the Earth is mainly nitrogen (78%) and oxygen (21%) with minor argon (1%) and carbon dioxide (0.03%). Its clear oceans include little salt (3.5%) and traces of sodium nitrate. The Earth's oxygen anomaly is related to the abundance of photosynthetic organisms, such as pond scum and trees. Evidence from paleosols for the persistence of chemically reactive oxygen in the atmosphere for at least the past 2450 million years can be taken as evidence for oxygenic photo-synthesis deep into the past. The isotopic fractionation of organic carbon back 3800 million years is evidence of other forms of photosynthesis even further back in time. Earth's cosmically peculiar atmosphere was produced and maintained by life. This is the core concept of Lovelock's Gaia hypothesis.

A prediction of the Gaia hypothesis is that life should have mechanisms for regulating such environmental features as atmospheric composition and temperature. On the mechanisms of regulation Lovelock was unclear. One of his ideas was that dimethyl sulfide gas produced by oceanic algae could seed clouds and act as a greenhouse gas to offset the algal consumption of carbon dioxide by photosynthesis. Later research has shown that dimethyl sulfide is indeed created by algae, but it is not an abundant or influential gas in our atmosphere.

I think Lovelock should have looked closer to home, to the fields around the Cornish mill that he has converted to a home and laboratory. According to the Proserpina Principle, soil is a great regulator of Earth's temperature, consuming carbon dioxide to produce life and clay, and releasing carbon dioxide from fire and oxidation of organic matter during erosion. The balance of producers and consumers in soil keeps these gases in check on a variety of time scales: days, years, millennia, and eons. Producers and consumers are comparable with white and black daises in Lovelock's favorite Daisyworld analogy. Imagine a world in which white daises reflect sunlight and cool the world, but black daises absorb light and warm the world. The mix of white and black daises would form its own balance of nature to regulate this imagi-nary planet's temperature. If external inputs of sunlight diminish, black daises will increase, and if temperatures climb, white daises will increase. Daisy-world is of course fanciful, unlike the balance of producers and consumers, and the mosaic of different soil types. We have seen that some kinds of soils such as Mollisols, Ultisols and Oxisols support high productivity vege-tation that is a force for global cooling like white daises. On the other hand, Gelisols and Aridisols support low productivity taiga and desert vegetation that cannot offset the continuing emission of carbon dioxide from volca-noes, somewhat like black daises. During greenhouse spikes of the past to more than 2000 ppm, Oxisols and Ultisols have spread to polar regions and Mollisols into deserts to curb the unusually high carbon dioxide. During glaciations, Gelisols and Aridisols were widespread and set a lower bound of atmospheric carbon dioxide at about 180 ppm. Gelisols and Aridisols are bright with high albedo, so cannot be an albedo thermostat because they would promote runaway cooling by expanding during glacial periods. Conversely, forests and grasslands are less reflective and their spread during greenhouse spikes would be a force for runaway warming if albedo were the only consideration. Oxisols, Ultisols and Mollisols consume more carbon as biomass and during weathering than Gelisols and Aridosols, and so can regulate extremes of warmth as a true thermostat. Little carbon is consumed by Gelisols and Aridisols, but their limited carbon sequestration is eventually overwhelmed by continuing degassing of carbon dioxide from volcanoes and springs. Soils have thus been a global thermostat preventing Earth from irreparable freezing like Mars, or terminal heating like Venus.

Lovelock's Gaia hypothesis did not meet with universal acceptance among scientists. On this score it is in good company with other hypotheses such as the Darwin-Wallace theory of evolution and Walter Alvarez's idea of asteroid impact at the end of the Cretaceous. A part of Gaia's problem was her mythopoetic name, suggested to Lovelock by his friend William Golding, British author of "Lord of the Flies." Mythic and holistic, Gaia seemed to be more akin to astrology, tarot cards, numerology and cheirology, than a serious scientific theory. The growth in popularity of great goddess mysticism did little to win over skeptical scientists.

An early criticism was the teleological argument that there is nothing unusual about the way things are, because that is how they were made and remain. Seamless interaction of organisms and environment is a common appearance of successfully coevolved ecosystems. The clue that evolution rather than special creation has been involved is minor imperfections, such as the gills and tails of human embryos, revealing our evolutionary kinship with fish. The Goldilocks view of Venus as too hot, Mars as too cold, and Earth as just right could be seen as consequences of their varied distance from the sun, and this in turn as a matter of chance, or by some as divine providence. But it is not quite this simple, as I have tried to show from the fossil record of soils on each of these planets. The atmospheres of Mars and Venus are at lifeless chemical equilibrium, and Mars at least, has landscapes and soils little changed for the past 4000 million years. Earth on the other hand, has seen marked fluctuation in atmospheric gas composition paced with innovations in the evolution of life and soils. The carbon dioxide content of our atmosphere has fluctuated considerably on time scales of hundreds of millions of years, as Ordovician and Jurassic greenhouses alternated with Carboniferous and Pleistocene icehouses, and as millipedes, trees, termites and grasses came to dominate terrestrial ecosystems. During the past 1.6 million years, carbon dioxide levels have fluctuated by lesser amounts on time scales of hundreds of thousands of years as ice caps and deserts have advanced and retreated over the continents of the northern hemisphere. On annual time scales, atmospheric carbon dioxide rises slightly with soil respiration and leaf fall in the northern hemisphere, and carbon dioxide levels fall by a comparable amount as trees come back into leaf and productivity in the spring and summer. The rhythms of carbon dioxide abundance correspond to biological activity on a variety of time scales. Just as organisms have physiological rhythms, such as heart beat, vulnerable to disruption, so do minor fluctuations in atmospheric gases reveal the inner workings of complex systems of feedback between organisms and their environments. The Proserpina Principle is an atmospheric component of planetary physiology. Fluctuation in Earth's atmospheric composition shows that our planet is alive and kicking, not frozen or cremated in static chemical equilibrium like the atmospheres of Mars and Venus.

Others criticized Gaia as unfalsifiable, because no alternative hypothesis was presented. There is an alternative view of the evolution of the Earth, which I have called Ereban, after the primeval darkness of Greek myth. According to the Ereban view, life arose by the most remarkable accidents only on those planets that had a narrow range of physicochemical conditions to allow it. Life has persisted in a generally hostile environment only where specific activities and abilities or organisms proved advantageous for scraping a living from available resources. In the Ereban view, the pervasive influence of life is illusory because life still depends ultimately on gases, liquids and rocks erupted as a byproduct of the internal differentiation of the Earth and on the quality and quantity of radiation from the Sun. The thin rind of the biosphere has been severely curtailed by impacts of large asteroids. Evidence for volcanic eruptions, meteorite impacts, and mass extinctions in the geological past demonstrate that Ereban hypotheses need to be taken seriously.

The Ereban view of life's history as a heroic struggle can be found in many geological textbooks and popular accounts of Earth history. It is Joseph Campbell's familiar "hero's journey", still recycled in movies such as "Star Wars". By this way of thinking, life evolved by incredibly long odds in the primordial soup of the world ocean or the toxic gush of a deep-sea hydrothermal vent. The earliest life consumed preformed organic molecules of the primeval soup and so was limited by the availability of these inorganically synthesized materials. When photosynthetic organisms evolved, they were mercilessly irradiated by near-lethal ultraviolet radiation unfiltered by the ozone layer. They also struggled to obtain nutrients such as phosphorus in the sea. They fought off predation by ever more sophisticated consumers. Oxygen released by these struggling early photosynthesizers was initially scavenged by chemically reduced iron minerals and decaying organic compounds. It was not until these sinks for oxygen were buried and photosynthetic activity increased in extensive shallow seas of continental margins that oxygenation of the atmosphere became noticeable. After some time, oxygen accumulated to the extent that its byproduct ozone in the stratosphere, 6–30 miles up, filtered out harmful shortwave radiation. Only then could life colonize formerly barren land surfaces. Photosynthetic ecosystems conquered formerly hostile habitats while poisoning with oxygen an early fermentative ecosystem that lingers on in swamps, stagnant oceans, and the guts of animals. By the Ereban view of life, internecine warfare over scarce resources between different branches of the burgeoning family tree of life drove other evolutionary innovations. Early invertebrates took up armor against their predators in the form of the first sea shells. A great armada of plants invaded the land from the sea once they had become armed with such adaptations as cuticles, stomates and spores. Trees prevailed over primitive land plants by poisoning competitors with toxins and shading them from the sun's energy. Trees wilted in dusty dry deserts created by global cooling and drying of the ice ages, but grasses were able to endure these hostile habitats. Volcanic eruptions and large meteorite impacts episodically wrought havoc on these struggling ecosystems, reasserting the primacy of raw physical power. On millennial timescales, organisms are at the mercy of wobbles in the Earth's orbit periodically perturbing the climate system so that glaciers scrape clean of life large areas of the continents. This is nature red in tooth and claw, a war of all against all. Large impact craters, voluminous volcanic ashes and thick flood deposits provide impressive confirmation of Ereban, or nonbiological, forcings in the geological past.

The Gaian view and its Proserpinan mechanism in contrast, are that life is a self-stabilizing surficial system. An acidic early atmosphere, rich in carbon dioxide and water, was neutralized by the formation of clayey soils. These became more effective at holding the landscape together as organic matter was created by photo-oxidation and other prebiotic syntheses of organic matter. In the complex, reactive, alternately wet and dry cracks and cavities of stable clayey soils, organic matter accumulated to produce soils similar to carbonaceous chondritic meteorites. Increasingly complex organic compounds were preserved as long as they promoted the persistence of soil against erosion. Organic matter, microbial scums, liverwort polsters and tropical rain forests can be seen as increasingly effective ways of stabilizing the landscape. From a home base rooted in nutrient cycling on land, life spread into the sea, the deep ocean and even into the air. An early planetary engineering feat of life was oxygenation of the atmosphere by cyanobacterial photosynthesis and weathering processes. Carbon dioxide is not only a fuel for photosynthetic production of organic matter, but forms carbonic acid active in leaching

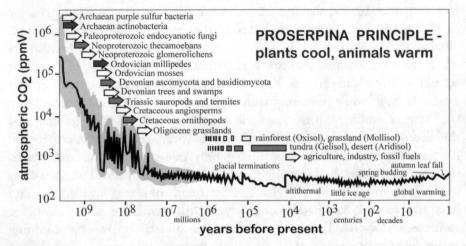

Fig. 16.4 Variation in carbon dioxide content of the atmosphere over the history of the Earth on a logarithmic time scale, which shows eons to the left and days to the right. The evolutionary advent of major groups of animals at times of greenhouse paleoclimate and of major groups of plants at times of icehouse is a reflection of the Proserpina Principle acting on long geological time scales

mineral nutrients from the soil. When consumers of the early photosynthesizers evolved, they contributed carbon dioxide to the air and thus a warmer temperature of the planet. Some microbial consumers not only oxidized carbon within organisms, but also carbon buried within sediments and soils. The Gaian view is one of planetary symbiosis, but it left unspecified particular mechanisms of control such as the Proserpina Principle. As we have seen, the advent of microbial consumers such as fungi and actinobacteria, of invertebrate consumers such as millipedes, and of tree-eaters such as dinosaurs and termites ushered in greenhouse eras in Earth history. The evolution of indigestible materials by photosynthesizers, such as the chitin of lichens, lignin of trees, and phytoliths of grasses, tipped the balance more toward carbon sequestration, thus ushering in ice ages. On the other hand evolution of the ability of cells to eat other cells, of jaws, and of high crowned grazing teeth opened new opportunities for warming as plants could be consumed more efficiently. This is like the annual variation on atmospheric carbon dioxide of the Proserpina Principle, but on long time scales (Fig. 16.4). It is a Gaian thermostat, revealed in part from the fossil record of soils. Rather than a cut-throat Ereban struggle, the Gaian view of life on Earth was a process of learning to do things better.

The cycles of soil and atmospheric fluctuation here called the Proserpina Principle have helped withstand Ereban forcings. Soils and ecosystems do not have the finely tuned thermostatic effect originally envisaged by Lovelock's

Gaia hypothesis. There are distinct oscillations of greenhouse gas abundance on daily, annual, millennial and eonal time scales. Stable oscillating systems are common in nature. For example, records of furs received by the Hudson Bay Company during the late nineteenth century show marked peaks in abundance of lynx pelts, alternating with abundance of the skins of snowshoe hares. Predators became abundant as prey declined in number, and vice versa. The wavelength of this oscillation at 9.7 years, does not seem to be related to sunspot or other astronomical cycles, but is an intrinsic period of population biology related to growth and reproduction rates of lynx and hare. These intrinsic fluctuations are built-in disturbances, which give the ecosystem great resilience in the face of extrinsic forcings, such as the unreasonably high level of trapping by Europeans. Similarly the cycles of greenhouse gas emission and sequestration in ecosystems and soils may have enabled Earth to survive large asteroid impact and enormous hydrocarbon degassing events. Greenhouse spikes led to geographic expansion of tropical forest and grassland soils with carbon sequestration sufficient to restore equilibrium. Spread of low productivity tundra and deserts during ice ages is eventually offset by continued emission of carbon dioxide from volcanoes and springs. Natural oscillations allow life to roll with bigger punches.

In traditional Chinese philosophy and medicine, a healthy life has balance between the masculine principle or yang and the female principle or yin. War and struggle allow us to grow, as do care and feeding. The great disasters of Earth history have taken a toll, but they also shaped the future. Dinosaurs are now gone, but mammals would not have had a chance without their passing. Quietly and relentlessly plants and soils covered and modified impact craters and volcanic tuffs. The scale of an interplanetary collision able to completely extinguish soil formation would be awesome indeed. Wider forces of stellar evolution can bring an end to soil formation as we know it, as attested by the hellish greenhouse of Venus and barren icehouse of Mars. In the end, hundreds of millions of years hence, when the sun expands to swallow up Venus, soil formation and life as we know it will not be possible on Earth either. But between the beginning and the end life has grown and diversified, gaining control and tenacity at a rate that exceeded the onslaught of incoming destruction. The meadow grows into a forest, which can withstand a flood that lays waste to the meadow nearby, eroding it into gullies and burying it in silt. The forest dampness, shade and stillness create a home for other creatures unsuited to hot dusty clearings. Eventually there is a storm or extraterrestrial impact so great that the forest dies by erosion or burial. As the ecosystem has developed, so has the soil that fed and supported it. Although still, dark and quiet, soils and paleosols reveal much about that vital time between the beginning and end of plant communities.

Soils and life share many qualities. Like organisms, soils are a complex mix of organic and mineral, flesh and bone, wood and leaf, crust and scum. Each soil is as individual as a person, yet there are recognizably different kinds of soils. Soils like people are controlled by and exert controls on their environment. They take in and give off gases, and regulate and consume fluids. Even as they grow in depth and degree of weathering, soils accumulate the waste of some creatures as the nutrition of others. The rhythms of annual leaf fall, of decadal predator and prey fluctuations, and of millennial nutrient depletion and renewal are like the dance of muscle and nerve that create the rhythm of a heart-beat. Like us also, soils and ecosystems have become better at what they do over geological time: they have learned. The speed and efficiency of weathering in a grassland soil of the late Miocene or a forest soil of the Devonian, compared with microbial soils of the Precambrian, is like the difference between computers and typewriters versus Medieval hand-copying of manuscripts. Like soil, we also live by consumption and production in the zone of materials and energy transfer. Soils also meet untimely deaths, buried by thick flood alluvium or eroded from their underlying rock. Eventually, soils run out of mineral nutrients or must cope with newly evolved or introduced forms of life, leaving them vulnerable to destruction. Like organisms, soils also have a fossil record and include a record of fossilized land creatures. The fossil record of soils informs us of their fundamental role in the deep geological past. In some ways, life can be considered soil grown tall, freed but not entirely liberated from the shackles of its mineral underpinnings.

Although Lovelock considered the whole Earth a goddess Gaia, and soils her living skin, by that view she may be alone in the universe, or at least very isolated. His metaphor breaks down when evolution of Earth's surface environments is considered. Can a creature evolve in isolation? What would select for against a particular kind of Earth by that view? I prefer a vision of soils as goddesses, who like Proserpina, change with seasonal and other rhythms of life. These goddesses are as varied and numerous as the ecosystems they support. The Histosols of cypress swamps, Alfisols of oak forest and Mollisols of the prairie all have distinct roles in a global economy of air, water, and nutrient elements. They all have different human uses and potential. We should stop taking them for granted, and learn to love them for what they are. If we destroy the soils that nurtured us this far, our feet will have merely so much clay. The problems of soil erosion and acidification are large and serious. Pessimism about the ongoing loss of topsoil and fertility can seem at times, like the bitter cold rain and mire that keeps a mythic hero from his quest. But there is an alternative view of soils and life. One person can make a difference, by cultivating and restoring one small garden of soil. Our understanding and care of soil can fall like the warm spring rain on the ground of human consciousness. As the drops sink into the rich brown earth and its bounty of seeds, blades of green will appear to transform the world once more. Proserpina returns.

MOTHER EARTH

Like a rose amid a charred black ruin

Of war-torn London or Berlin,

From swirling rocks in darkness,

A chaos of primeval Erebus,

A blue and living orb spun forth

Soil goddesses of mother Earth.

Divine soils filter water and air,

Bind the hills with grace and care.

Erebus strikes with volcanic plume,

And asteroid impact spreading doom.

But life has grown since it began

In tilted cycles, more yin than yang

Further Reading

Ager DV (1973) The nature of the stratigraphical record. Wiley, New York, p 114

Alvarez W (1997) T. rex and the crater of doom. Princeton University Press, Princeton, 185 p

Amundson R, Harden J, Singer M (eds) (1994) Factors in soil formation—a fiftieth anniversary perspective, vol 33. Special Publication of the Soil Science Society of America, Madison, 160 pp

Andersen GG, Borns HW (1994) The ice age world. Oxford University Press, New York, p 208

Baker VR (1982) The channels of Mars. University of Texas Press, Austin, p 198

Bakker RT (1986) The dinosaur heresies: new theories unlocking the mystery of the dinosaurs and their extinction. William Morrow and Company, New York, p 481

Barthel KW, Swinburne NHM, Morris SC (1978) Solnhofen: a study in Mesozoic palaeontology. Cambridge University Press, Cambridge, p 236

Basu AR, Petaev MI, Poreda RJ, Jacobsen SB, Becker L (2003) Chondritic meteorite fragments associated with the Permian-Triassic boundary in Antarctica. Science 302:1388–1392

Bear FE (1986) Earth: the stuff of life (revised by H. W. Pritchard and W. E. Akin). University of Oklahoma Press, Norman, p 318

Bernal JD (1967) The origin of life. (includes translation of article in Russian by I.A.Oparin). World, New York, p 345

Bobrovskiy I, Hope JM, Ivantsov A, Nettersheim BJ, Hallmann C, Brocks JJ (2018) Ancient steroids establish the Ediacaran fossil *Dickinsonia* as one of the earliest animals. Science 361:1246–1249

Bolen JS (1984) Goddesses in everywoman. Harper and Row, New York, p 334

Bragg M (1983) Land of the lakes. Secker and Warking, London, p 249

G. J. Retallack, *Soil Grown Tall*,
https://doi.org/10.1007/978-3-030-88739-1

Brasier MD, Green OR, Jephcoat AP, Kleppe AK, Van Kranendonk MJ, Lindsay JF, Stale A, Grassineau NV (2002) Questioning the evidence for Earth's oldest fossils. Nature 416:76–81

Brunet M, Guy F, Mackaye HT, Likius A, Ahounta D, Beauvilain A, Blondel C, Bocherens H, Boisserie JR, de Bonis LD, Coppens Y, Dejax J, Denys C, Ruringer P, Eisenmann V, Fanone G, Frouty P, Geraards J, Lehmann T, Lihoreau F, Luochart A, Mahamout A, Merceron G, Mouchelin G, Otero O, Campomanes PP, de Leon MP, Rage JC, Sapanet M, Schuster M, Sudre J, Tassy P, Valentin X, Vignaud P, Viriot L, Zazzo A, Zollikofer C (2002) A new homnid from the upper Miocene of Chad, central Africa. Nature 418:145–148

Bullfinch T (1985) The golden age. Reprinted by Bracken Books. Studio Editions, London, p 495

Burkert W (1985) Greek religion (Translated by John Raffan). Harvard University Press, Cambridge, Massachusetts, 493 pp

Cairns-Smith AG (1971) The life puzzle. University of Toronto Press, Toronto, p 165

Cairns-Smith AG (1982) Genetic takeover. Cambridge University Press, Cambridge, p 427

Cairns-Smith AG, Hartman H (eds) (1986) Clay minerals and the origin of life. Cambridge University Press, Cambridge, p 193

Campbell IB, Claridge GGC (1987) Antarctica: soils, weathering processes and environment. Developments in soil Science no 16. Elsevier, Amsterdam, 368 pp

Carpenter R (1946) Folk tale, fiction and saga in the Homeric epics. University of California Press, Berkeley, p 198

Carr MH (1981) The surface of Mars. Yale University Press, New Haven, p 232

Cavosie AJ, Valley JW, Wilde SA (2007) The oldest terrestrial mineral record: a review of 4400 to 4000 Ma detrital zircons from Jack Hills, Western Australia. Precambr Geol 15:91–111

Chaloner WG, Lawson JD (eds) (1985) Evolution and environment in the Late Silurian and Early Devonian. Phil Trans R Soc Lond Biol Sci B309:342

Colbert EH (1995) The little dinosaurs of Ghost Ranch. Columbia University Press, New York, p 247

Cone J (1991) Fire under the sea: the discovery of the most extraordinary environment on Earth—volcanic hot springs on the ocean floor. William Morrow, New York, p 285

Cotterell A (1989) The Macmillan illustrated encyclopedia of myths and legends. Macmillan, New York, p 260

Crane PR, Friis EM, Pedersen KJ (1995) The origin and early diversification of angiosperms. Nature 374:27–33

Crick FHC (1981) Life itself: its origin and nature. Touchstone Press, Simon and Schuster, New York, p 192

Czerkas SJ, Czerkas SA (1996) Dinosaurs: a global view. Barnes and Noble Boks, New York, p 247

Dilcher DL, Crane PR (1984) In pursuit of the first flower. Nat Hist 93(3):57–61

Driese SG, Nordt LE (eds) New frontiers in paleopedology and terrestrial paleoclimatology. Society of Economic Paleontologists and Mineralogists Special Paper, vol 44, 5–16

Eisler R (1988) The chalice and the blade: our history, our future. Harper and Row, San Francisco, p 261

Erlandson JM, Graham MH, Bourque BJ, Corbett D, Estes JA, Steneck RS (2007) The kelp highway hypothesis: marine ecology, the coastal migration theory, and the peopling of the Americas. J Island Coast Archaeol 2:161–174

Erwin DH (1993) The great Paleozoic crisis: life and death in the Permian. Columbia University Press, New York, p 327

Evans SD, Droser ML, Gehling JG (2015) *Dickinsonia* liftoff: evidence of current derived morphologies. Palaeogeogr Palaeoclimatol Palaeoecol 434:28–33

Fitzpatrick EA (1980) Soils. Longman, London, p 353

Flannery TF (1994) The future eaters: an ecological history of the Australasian lands and people. Reed-New Holland, Sydney, p 425

Fleagle JG (1988) Primate adaptation and evolution. Academic Press, Harcourt Brace Jovanovich, San Diego, p 486

Frakes LA, Francis J, Syktus JI (1992) Climate modes of the Phanerozoic: the history of Earth's climate over the past 600 million years. Cambridge University Press, Cambridge, p 274

Friis EM, Pedersen KR, Crane PR (2006) Cretaceous angiosperm flowers: innovation and evolution in plant reproduction. Palaeogeogr Palaeoclimatol Palaeoecol 232:251–293

Frymer-Kensky T (1992) In the wake of the goddess: women, culture, and the biblical transformation of pagan myth. Free Press, New York, p 292

Gensel PG, Andrews HN (1984) Devonian paleobotany. Praeger Press, New York, p 381

Gensel PG, Edwards D (eds) (2001) Plants invade the land: evolutionary and environmental perspectives. Columbia University Press, New York, p 304

Gimbutas M (1989) The language of the goddess. Harper and Row, San Francisco, p 388

Glaessner MF (1984) The dawn of animal life. Cambridge University Press, Cambridge, p 244

Glenn W (ed) (1994) The mass extinction debates: how science works in a crisis. Stanford University Press, Stanford, p 370

Gore A (1992) Earth in balance. Plume (Penguin), New York, p 407

Greeley R (1985) Planetary landscapes. George Allen and Unwin, London, p 265

Grootes PM, Stuiver M, White JWC, Johnsen S, Jouzel J (1993) Comparison of oxygen isotope records from the GISP3 and GRIP Greenland ice cores. Nature 366:552–554

Haile-Selassie Y (2001) Miocene hominids from the middle Awash, Ethiopia. Nature 412:178–181

Haldane JBS (1929) The origin of life. Rationalist Ann 148:3–10

Hale ME (1983) The biology of lichens. Edward Arnold, London, p 190

Hallam A, Wignall PB (1997) Mass extinctions and their aftermath. Oxford University Press, New York, p 320

Hay RL (1976) Geology of Olduvai Gorge. University of California Press, Berkeley, p 203

Heywood VH (1978) Flowering plants of the world. Oxford University Press, Oxford, p 335

Hillel DJ (1991) Out of the earth. Free Press, New York, p 321

Holland HD (1984) The chemical evolution of the atmosphere and ocean. Princeton University Press, Princeton, p 582

Hoyt WG (1987) Coon Mountain controversies: meteor Crater and the development of impact theory. University of Arizona Press, Tucson, p 370

Jacobs B, Kingston JD, Jacobs L (1999) The origin of grass-dominated ecosystems. Ann Mo Bot Gard 86:673–696

Jenkins DL, Davis LG, Stafford TW, Campos PF, Hockett B, Jones GT, Cummings LS, Yost C, Connolly TJ, Yohe RM, Gibbons SC, Raghavan M, Rasmussen M, Paijmans JLA, Hofreiter M, Kemp BM, Barta JL, Monroe C, Gilbert MTP, Willerslev E (2012) Clovis age western stemmed projectile points and human coprolites at the Paisley Caves. Science 337:223–228

Jenny H (1994) Factors of soil formation: a system of quantitative pedology. (Reprint of the classic work of 1941). Dover Publications, New York, p 281

Johanson DC, O'Farrell K (1990) Journey from the dawn: life with the world's first family. Villard Books, New York, p 123

Johanson D, Edgar B, Brill D (1996) From Lucy to language. Simon and Schuster, New York, p 272

Joseph LE (1991) Gaia: the growth of an idea. St Martins Press, New York, p 276

Jowett B (1937) The dialogues of Plato, vol 2. Random House, New York, 939 pp

Karner DB, Muller RA (2000) A causality problem for Milankovitch. Science 288:2143–2144

Knoll AH (2003) Life on a young planet. Princeton University Press, Princeton, p 277

Kramer SN (1944) Sumerian mythology. Mem Am Phil Soc 21:25

Krull ES, Retallack GJ (2000) $\delta^{13}C$ depth profiles from paleosols across the Permian-Triassic boundary: evidence for methane release. Geol Soc Am Bull 112:1459–1472

Kühnelt W (1976) Soil biology, 2nd edn. Michigan State University Press, East Lansing, p 483

Levi-Setti R (1993) Trilobites, 2nd edn. University of Chicago Press, Chicago, p 342

Lockley M (1991) Tracking dinosaurs. Cambridge University Press, Cambridge, p 238

Logan WB (1995) Dirt: the ecstatic skin of the Earth. Riverhead Books, New York, p 202

Lovelock JE (1979) Gaia: a new look at life on Earth. Oxford University Press, Oxford, p 157

Lovelock JE (1988) The ages of Gaia: a biography of our living Earth. W. W. Norton, New York, 252 pp

Lovelock JE (2000) Gaia: the practical science of planetary medicine. Gaia Press, Sroud, p 192

Lowe JJ, Walker MJC (1997) Reconstructing Quaternary environments, 2nd edn. Longman, London, p 446

MacFadden BJ (1992) Fossil horses: systematics, paleobiology and evolution of the family Equidae. Cambridge University Press, Cambridge, p 369

Manchester SR (1994) Fruits and seeds of the middle Eocene Nut Beds flora, Clarno Formation, Oregon. Paleontographica Americana 58:205 p

Margulis L (1981) Symbiosis and cell evolution. W.H. Freeman, San Francisco, p 419

McFarlane MJ (1976) Laterite and landscape. Academic Press, New York, p 151

McKay DS, Gibson EK, Thomas-Kepeta KL, Vali H, Romanek CS, Clemett SJ, Chillier XDF, Maechling CR, Zare RN (1996) Search for past life on Mars: possible relic biogenic activity in Martian meteorite ALH84001. Science 273:924–930

McKenzie N, Jacquier D, Isbell R, Brown K (2004) Australian soils and landscapes. CSIRO Press, Melbourne, p 416

McKerrow WS (1978) The ecology of fossils. M.I.T. Press, Chicago, p 213

McMenamin MAS, McMenamin DLS (1990) The emergence of animals: the Cambrian breakthrough. Columbia University Press, New York, p 217

McSween HY (1987) Meteorites and their parent planets. Cambridge University Press, Cambridge, p 233

McSween HY (1997) Evidence for life in a Martian meteorite? GSA Today 7(7):1–7

Miller MF, Hasiotis ST, Babcock LE, Isbell JL, Collinson JW (2001) Tetrapod and large burrows of uncertain origin in Triassic high paleolatitude floodplain deposits, Antarctica. Palaios 16:218–232

Monod J (1971) *Chance and necessity.* (Translated by A. Wainhouse). Knopf, New York, p 199

Nisbet EG (1987) The young Earth. George Allen and Unwin, Boston, p 402

Nutman AP, Bennett VC, Friend CR, Van Kranendonk MJ, Chivas AR (2016) Rapid emergence of life shown by discovery of 3700 million-year-old microbial structures. Nature 537:535–538

Ollier C (1991) Ancient landforms. Belhaven Press, London, p 233

Olsen PE (1993) The terrestrial plant and herbivore arms race: a major control of Phanerozoic CO_2? Abstr Geol Soc Am 25(3):71

Paepe R, Van Overloop E (1990) River and soils cyclicities interfering with sea level changes. In: Paepe R, Fairbridge RW, Jelgersma S (eds) Greenhouse effect, sea level and mitigation of drought. Kluwer Academic, Boston, pp 253–280

Paton TR, Humphreys GS, Mitchell PB (1995) Soils: a global view. UCL, London, p 213

Peterson KJ, Waggoner B, Hagadorn JW (2003) A fungal analog for Newfoundland Ediacaran fossils? Integr Comp Biol 43:127–136

Pflug H (1973) Zur fauna der Nama-Schichten in Südwest-Afrika IV. Mikroscopische Anatomie Der Petalo-Organisme. Palaeontographica B144:144–166

Pickett JW (2003) Stratigraphic relationships of laterite at Little Bay, near Maroubra, New South Wales. Aust J Earth Sci 50:63–68

Pickford M (1975) Late Miocene sediments and fossils from the northern Kenya Rift Valley. Nature 256:279–284

Pimentel D, Harvey C, Resosudarmo P, Sinclair K, Kurz D, McNair M, Crist S, Shpritz L, Fitton L, Saffouri R, Blair R (1995) Environmental and economic costs of soil erosion and conservation benefits. Science 267:1117–1122

Prothero DR (1994) The Eocene-Oligocene transition: paradise lost. Columbia University Press, New York, p 291

Radosevich SC, Retallack GJ, Taieb M (1992) A reassessment of the paleoenvironment and preservation of hominid fossils from Hadar, Ethiopia. Am J Phys Anthropol 87:15–27

Reinhardt J, Sigleo WR (eds) (1988) Paleosols and weathering through geologic time: principles and applications, vol 216. Special Paper of the Geological Society of America, Boulder, 181 pp

Retallack GJ (1991) Miocene paleosols and ape habitats in Pakistan and Kenya. Oxford University Press, New York, p 346

Retallack GJ (1991) Untangling the effects of burial alteration and ancient soil formation. Annu Rev Earth Planet Sci 19:183–206

Retallack GJ (1992) Middle Miocene fossil grasses from Fort Ternan (Kenya) and evolution of African grasslands. Paleobiology 18:383–400

Retallack GJ (1992) What to call early plant formations on land. Palaios 7:508–520

Retallack GJ (1994) A pedotype approach to latest Cretaceous and earliest Tertiary paleosols in eastern Montana. Bull Geol Soc Am 106:1377–1397

Retallack GJ (1994) Were the Ediacaran fossils lichens? Paleobiology 20:523–544

Retallack GJ (1995) Pennsylvanian vegetation and soils. In: Cecil B, Edgar T (eds) Predictive stratigraphic analysis, vol 2110. Bulletin of the U.S. Geological Survey, p 13–19

Retallack GJ (1995) Permian-Triassic life crisis on land. Science 267:77–80

Retallack GJ (1996) Acid trauma at the Cretaceous-Tertiary boundary in eastern Montana. GSA Today 6(5):1–7

Retallack GJ (1996) Early Triassic therapsid footprints from the Sydney Basin. Alcheringa 20:301–314

Retallack GJ (1996) Paleosols: record and engine of past global change. Geotimes 41(6):25–28

Retallack GJ (1997) A colour guide to paleosols. Wiley, Chichester, p 175

Retallack GJ (1997) Dinosaurs and dirt. In: Wolberg D, Stump E (eds) Dinofest II. Philadelphia Academy of Sciences, Philadelphia, pp 345–359

Retallack GJ (1997) Early forest soils and their role in Devonian global change. Science 276:583–585

Retallack GJ (1997) Neogene expansion of the North American prairie. Palaios 12:380–390

Retallack GJ (1997) Palaeosols in the upper Narrabeen Group of New South Wales as evidence of Early Triassic palaeoenvironments without exact modern analogues. Aust J Earth Sci 44:185–201

Retallack GJ (1999) Postapocalyptic greenhouse paleoclimate revealed by earliest Triassic paleosols in the Sydney Basin, Australia. Geol Soc Am Bull 111:52–70

Retallack GJ (2000) Ordovician life on land and early Paleozoic global change. In: Gastaldo RA, DiMichele, WA (eds) Phanerozoic terrestrial ecosystems. Paleontological Society Short Course Notes (in press)

Retallack GJ (2001) A 300 million year record of atmospheric CO_2 from fossil plant cuticles. Nature 411:287–290

Retallack GJ (2001) Cenozoic expansion of grasslands and global cooling. J Geol 109:407–426

Retallack GJ (2001) *Scoyenia* burrows from Ordovician paleosols of the Juniata Formation in Pennsylvania. Palaeontology 44:209–235

Retallack GJ (2004) End-Cretaceous acid rain as a selective extinction mechanism between birds and dinosaurs. In: Currie PJ, Koppelhus EB, Shugar MA, Wright JL (eds) Feathered dragons: studies on the transition from dinosaurs to birds. Indiana University Press, Bloomington and Indianapolis, p 35–64

Retallack GJ (2004) Soils and global change in the carbon cycle over geological time. In: Drever JI (ed) Surface and groundwater, weathering and soils, vol 5. In: Holland HD, Turekian KK (eds) Treatise of geochemistry. Alsevier, Amsterdam, p 581–605

Retallack GJ (2011) Woodland hypothesis for Devonian evolution of tetrapods. J Geol 119:235–258

Retallack GJ (2013) Ediacaran life on land. Nature 493:89–92

Retallack GJ (2013) Permian and Triassic greenhouse crises. Gondwana Res 24:90–103

Retallack GJ (2014) Paleosols. In: Henke W, Tattersall I (eds) Handbook of paleoanthropology, vol 1. Principles, methods and approaches. Springer, Berlin, p 383–408

Retallack GJ (2014) Precambrian life on land. Palaeobotanist 63:1–15

Retallack GJ (2014) Volcanosedimentary paleoenvironments of Ediacaran fossils in Newfoundland. Geol Soc Am Bull 126.619–638

Retallack GJ (2015) Acritarch evidence of a late Precambrian adaptive radiation of Fungi. Botanica Pacifica 4:19–33

Retallack GJ (2016) Astropedology: palaeosols and the origin of life. Geol Today 32:172–178

Retallack GJ (2016) Ediacaran fossils in petrographic thin sections. Alcheringa 40:583–600

Retallack GJ (2016) Ediacaran sedimentology and paleoecology of Newfoundland reconsidered. Sed Geol 333:15–31

Retallack GJ (2016) Field and laboratory tests for recognition of Ediacaran paleosols. Gondwana Res 36:94–110

Retallack GJ (2017) Exceptional preservation of soft-bodied Ediacara Biota promoted by silica-rich oceans: comment. Geology 44:e407

Retallack GJ (2018) *Dickinsonia* sterols not unique to animals. Science 361:1246. https://doi.org/10.1126/science.aat7228

Retallack GJ (2018) Interflag sandstone laminae, a novel fluvial sedimentary structure with implications for Ediacaran paleoenvironments. Sed Geol 379:60–76

Retallack GJ (2019) Ordovician land plants and fungi from Douglas Dam, Tennessee. Palaeobotanist 68:173–205

Retallack GJ (2019) Soils of the past. Wiley, Chichester, 534 p

Retallack GJ (2020) Boron paleosalinity proxy for deeply buried Paleozoic and Ediacaran fossils. Palaeogeogr Palaeoclimatol Palaeoecol 540:109536

Retallack GJ (2021) Ediacaran periglacial sedimentary structures. J Paleosciences. 70. https://blogs.uoregon.edu/gregr/wp-admin/post.php?post=58&action=edit

Retallack GJ (2021) Paleosols and weathering leading up to Snowball Earth in central Australia. Aust J Earth Sci 68. https://doi.org/10.1080/08120099.2021. 1906747

Retallack GJ, Dilcher DL (1981) A coastal hypothesis for the dispersal and rise to dominance of flowering plants. In: Niklas KJ (ed) Paleobotany, paleoecology and evolution. Praeger Press, New York, pp 27–77

Retallack GJ, Dilcher DL (1981) Early angiosperm reproduction: *Prisca reynoldsii* gen et sp. nov. from mid-Cretaceous coastal deposits in Kansas, U.S.A. Palaeontographica B179:103–137

Retallack GJ, Dilcher DL (1986) Cretaceous angiosperm invasion of North America. Cretac Res 7:227–252

Retallack GJ, Feakes CR (1987) Trace fossil evidence for Late Ordovician life on land. Science 235:61–63

Retallack GJ, Krinsley DH (1993) Metamorphic alteration of a Precambrian (2.2 Ga) paleosol from South Africa revealed by back-scatter imaging. Precambr Res 63:27–41

Retallack GJ, Germán-Heins J (1994) Evidence from paleosols for the geological antiquity of rain forest. Science 265:499–502

Retallack GJ, Mindszenty A (1994) Well preserved Late Precambrian paleosolsfrom northwest Scotland. J Sediment Res A64:264–281

Retallack GJ, Jahren AH (2008) Methane release from igneous intrusion of coal during Late Permian extinction events. J Geol 116:1–20

Retallack GJ, Huang C-M (2011) Ecology and evolution of Devonian trees in New York, USA. Palaeogeogr Paleoclimatol Paleoecol 299:110–128

Retallack GJ, Dilcher DL (2012) Core and geophysical logs versus outcrop for interpretation of Cretaceous paleosols in the Dakota Formation of Kansas. Palaeogeogr Palaeoclimatol Palaeoecol 329–330:47–63

Retallack GJ, Noffke N (2019) Are there ancient soils in the 3.7 Ga Isua Greenstone Belt, Greenland? Palaeogeogr Palaeoclimatol Palaeoecol 514:18–30

Retallack GJ, Broz A (2020) Ediacaran and Cambrian paleosols in central Australia. Palaeogeogr Palaeoclimatol Palaeoecol 560:110047

Retallack GJ, Conde GD (2020) Deep time perspective on rising atmospheric CO_2. Global Planet Change 189:e.103177

Retallack GJ, Dugas DP, Bestland EA (1990) Fossil soils and grasses of the earliest East African grasslands. Science 247:1325–1328

Retallack GJ, Bestland EA, Dugas DP (1995) Miocene paleosols and habitats of *Proconsul* on Rusinga Island, Kenya. J Hum Evol 29:53–91

Retallack GJ, Veevers JJ, Morante R (1996) Global early Triassic coal gap between Late Permian extinction and Middle Triassic recovery of peat-forming plants. Bull Geol Soc Am 108:195–207

Retallack GJ, Seyedolali A, Krull ES, Holser WT, Ambers CP, Kyte FT (1998) Search for evidence of impact at the Permian-Triassic boundary in Antarctica and Australia. Geology 26:979–982

Retallack GJ, Bestland EA, Fremd TR (2000) Eocene and Oligocene paleosols of central Oregon, vol 344. Geological Society of America Special Paper, 192 pp

Retallack GJ, Wynn JG, Benefit BR, McCrossin ML (2002) Paleosols and paleoenvironments of the middle Miocene Maboko Formation, Kenya. J Hum Evol (in press)

Retallack GJ, Krull ES, Thackray GD, Parkinson D (2013) Problematic urn-shaped fossils from a Paleoproterozoic (2.2 Ga) paleosol in South Africa. Precambr Res 235:71–87

Retallack GJ, Krinsley DH, Fischer R, Razink JJ, Langworthy K (2016) Archean coastal-plain paleosols and life on land. Gondwana Res 40:1–20

Richardson DHS (1981) The biology of mosses. Wiley, New York, p 220

Ruddiman WF (2003) The anthropogenic greenhouse era began thousands of years ago. Clim Change 61:261–293

Russell DA (1977) A vanished world: the dinosaurs of western Canada. Nat Mus Nat Sci, Ottawa, p 142

Russell DA (1989) An odyssey in time: the dinosaurs of North America. University of Toronto Press, Toronto, p 240

Russell P (1993) The Palaeolithic Mother-Goddess: fact or fiction? In: du Clos, H, Smith L (eds) Women and archaeology: a feminist critique, vol 23. Occasional Papers in Prehistory, Department of Prehistory, Research School of Pacific Studies, Australian National University, Canberra, pp 93–97

Ryder G, Fastovsky DE, Gartner G (eds) (1996) The Cretaceous-Tertiary boundary event and other catastrophes in Earth history. Special Publication of the Geological Society of America, Boulder, vol 307, 569 pp

Rye R, Holland HD (1998) Paleosols and the evolution of atmospheric oxygen: a critical review. Am J Sci 278:621–672

Savage RJG, Long R (1986) Mammal evolution. Facts on File, New York, p 259

Savory A, Butterfield J (1999) Holistic management: a new framework for decision making. Island Press, Washington, p 616

Schoene B, Samperton KM, Eddy MP, Keller G, Adatte T, Bowring SA, Khadri SF, Gertsch B (2015) U-Pb geochronology of the Deccan Traps and relation to the end-Cretaceous mass extinction. Science 347:182–184

Schopf JW (ed) (1983) Earth's earliest biosphere: its origin and evolution. Princeton University Press, Princeton, p 543

Schopf JW (1993) Microfossils of the early Archaean Apex Chert: new evidence of the antiquity of life. Science 260:640–646

Schopf JW (1999) Cradle of life: the discovery of Earth's earliest fossils. Princeton University Press, Princeton, p 367

Schopf JW, Klein C (eds) (1992) The Proterozoic biosphere: a multidisciplinary study. Cambridge University Press, Cambridge, p 1348

Schopf JW, Kudiyavtsev AR, Agresti DG, Wdowiak TJ, Czaja AD (2002) Laser Raman imagery of Earth's earliest fossils. Nature 416:73–76

Schwartzman D (1999) Life, temperature and the Earth. Columbia University Press, New York, p 241

Seilacher A (1989) Vendozoa: organismic construction in the Proterozoic biosphere. Lethaia 22:229–239

Seilacher A (1992) Vendobionta and Psammocorallia: lost constructions of Precambrian evolution. J Geol Soc London 149:607–613

Selley RC (1985) Ancient sedimentary environments, 3rd edn. Cornell University Press, Ithaca, p 317

Senut B, Pickford M, Gommery D, Mein P, Cheboi K, Coppens Y (2001) First hominid from the Miocene (Lukeino Formation, Kenya). Comptes Rendus Academie Des Sciences, Paris, Sciences De La Terre Et Planetes 332:137–144

Sharpton VL, Ward PD (eds) (1990) Global catastrophes in Earth history: an interdisciplinary conference on impacts, volcanism and mass mortality. Special Paper of the Geological Society of America (Boulder), vol 247, 631 p

Shipman P (1981) Life history of a fossil. Harvard University Press, Cambridge, p 222

Squyres SW, Arvidson RE, Bell JF, Brückner J, Cabrol NA, Calvin W, Carr MW, Christensen PR, Clark BC, Crumpler LR, Des Marais DJ, d'Uston C, Economou T, Farmer J, Farrand W, Folkner W, Golombek M, Gorevan SC, Grant JA, Greeley R, Grotzinger J, Haskin L, Herkenhoff KE, Hviid S, Johnson J, Klinghöfer G, Knoll AH, Landis G, Lemmon M, Li R, Madsen MB, Malin MC, McLennan SM, McSween HY, Ming DW, Moersch J, Morris RV, Parker T, Rice JW, Tichter L, Rieder R, Sims M, Smith M, Smith P, Soderblom LA, Sullivan R, Wänke H, Wdowiek T, Wolff M, Yen A (2004) The Opportunity Rover's Athena science investigation at Meridiani Planum, Mars. Science 306:1698–1703

Starhawk (1979) The spiral dance: rebirth of the ancient religion of the great goddess. Harper and Row, San Francisco, p 218

Steiner M, Reitner J (2001) Evidence of organic structures in Ediacara-type fossils and associated microbial mats. Geology 29:1119–1122

Sun G, Dilcher DL, Zheng S-L, Zhou Z-V (1998) In search of the first flower: a Jurassic angiosperm *Archaefructus*. Science 282:1692–1695

Tarhan LG, Droser ML, Gehling JG, Dzaugis MP (2017) Microbial mat sandwiches and other anactualistic sedimentary features of the Ediacara Member

(Rawnsley Quartzite, South Australia): implications for interpretation of the ediacaran sedimentary record. Palaios 32:181–194

Taylor DW, Hickey LJ (1990) An Aptian plant with attached leaves and flowers: implications for angiosperm origin. Science 247:702–704

Teague WR, Dowhower SL, Baker SA, Haileb N, DeLaunea PB, Conover DM (2011) Grazing management impacts on vegetation, soil biota and soil chemical, physical and hydrological properties in tall grass prairie. Agr Ecosyst Environ 141:310–322

Teague R, Provenza F, Kreuter U, Steffens T, Barnes M (2013) Multi-paddock grazing on rangelands: why the perceptual dichotomy between research results and rancher experience? J Environ Manage 128:699–717

The Diagram Group (1983) A field guide to dinosaurs. Avon, New York, p 256

Thomas K (1983) Man and the natural world. Pantheon, New York, p 426

Turek V, Marek J, Benes J, Brown J (1984) Fossils of the world. Arch Cape Press, New York, p 495

Van Andel TH, Runnels CN (1987) Beyond the Acropolis: a rural Greek past. Stanford University Press, Stanford, p 221

Vickers-Rich P, Rich TH (1993) Wildlife of Gondwana. Reed, Chatswood, Australia, 276 p

Volk T (1998) Gaia's body: towards a physiology of Earth. Copernicus Press, New York, p 269

Walker A, Shipman P (1996) The wisdom of the bones. A. A. Knopf, New York, 338 p

Wang H, Dilcher DL, Schwarzwalder RN, Kvaček J (2011) Vegetative and reproductive morphology of an extinct Early Cretaceous member of Platanaceae from the Braun's Ranch locality, Kansas, USA. Int J Plant Sci 172:139–157

Wang X, Zheng XT (2012) Reconsiderations on two characters of early angiosperm Archaefructus. Palaeoworld 21:193–201

Ward PD (1994) The end of evolution: on mass extinctions and the preservation of biodiversity. Bantam, New York, p 301

Watanabe Y, Stewart BW, Ohmoto H (2004) Organic- and carbonate-rich soil formation ~2.6 billion years ago at Schagen, East Transvaal district South Africa. Geochimica Et Cosmochimica Acta 68:2129–2151

White ME (1986) The greening of Gondwana. Reed, Frenchs Forest, Australia, p 256

Whittington HB (1985) The Burgess Shale. Yale University Press, New Haven, p 151

Williams GE, Schmidt PW (1997) Palaeomagnetic dating of sub-Torridon Group weathering profiles, NW Scotland: verification of Neoproterozoic paleosols. J Geol Soc London 154:987–997

Wilson HM, Anderson LI (2004) Morphology and taxonomy of Paleozoic millipedes (Diplopoda, Chilognatha, Archipolypoda) from Scotland. J Paleontol 78:169–184

Wing SL, Boucher LD (1998) Ecological aspects of the Cretaceous flowering plant radiation. Ann Rev Earth Planet Sci 26:379–421

Woodrow DL, Sevon WD (eds) (1985) The Catskill Delta. Special Paper of the Geological Society of America (Boulder), vol 201, 246 p

Wynn JG (2000) Paleosols, stable carbon isotopes and paleoenvironmental interpretation of Kanapoi, Northern Kenya. J Hum Evol 39:411–435

Wynn JG (2003) Miocene paleosols of Lothagam Hill. In Harris JM, Leakey MG (eds) Lothagam: the dawn of humanity in East Africa. Columbia University Press, New York, pp 31–42

Wynn JG, Retallack GJ (2001) Palaeoenvironmental reconstruction of middle Miocene paleosols bearing *Kenyapithecus* and *Victoriapithecus*. J Hum Evol 40:263–288

Yuan X, Xiao S, Taylor TN (2005) Lichen-like symbiosis 600 million years ago. Science 308:1017–1020

Index

© The Editor(s) (if applicable) and The Author(s), under exclusive license to Springer Nature Switzerland AG 2022
G. J. Retallack, *Soil Grown Tall*,
https://doi.org/10.1007/978-3-030-88739-1

Printed in the United States
by Baker & Taylor Publisher Services